Seven Bridges Road

Seven Bridges Road
From the Past to the Future
Kenneth W. Able

Seven Bridges Road: From the Past to the Future

Kenneth W. Able
Distinguished Professor Emeritus
Rutgers University Marine Field Station
132 Great Bay Blvd, Tuckerton, NJ 08087
Phone: (609) 214-2471

Published by South Jersey Culture & History Center
at Stockton University
101 Vera King Farris Dr.
Galloway, New Jersey 08205
United States
stockton.edu/sjchc/

Library of Congress Control Number: 2026911979

ISBN: 978-1-947889-34-7

STOCKTON UNIVERSITY
SOUTH JERSEY CULTURE & HISTORY CENTER

Table of Contents

Dedication

To the many volunteers who, over the last 35 years, have been central contributors to the research and education components of the Rutgers University Marine Field Station, including the Rutgers University Department of Marine and Coastal Sciences, Stockton University, and particularly the Jacques Cousteau National Estuarine Research Reserve. I am reminded of the significance of these contributions by one who epitomized what a volunteer could be. Steve Zeck died during the final stages of the preparation of this book.

Preface

I grew up interested in natural history. As a grade-schooler, when my family lived in the inner city of Indianapolis, very close to a railroad track, I managed to find standing puddles of water that intrigued me and that I fished in even though I knew no fish could live there. My view of the world changed dramatically when the family moved to the suburbs and I had access to a vegetable garden and a flowing creek with real fish, frogs, insects and plants. High school and college provided expanded access to the natural world with a scientific bent. But my eyes were really opened when I left Indiana for graduate school in Virginia and was exposed, for the first time, to an estuary, notably the Chesapeake Bay. From there, over the next several decades, I was involved in research and learning from the Gulf of St. Lawrence in Canada to the Gulf of Mexico, but my focus was always on my "backyard" after I moved to the Rutgers University of Marine Field Station in southern New Jersey.

There I found opportunities to observe and study a single estuary across the seasons, developing insights uniquely possible because of the location. From my office, which sits above the intertidal zone inside Little Egg Inlet on Great Bay, I can see bottlenose dolphins in the summer, harbor seals in the winter, and note the other seasonal changes in the adjacent marsh. I can do this during king tides, sunny day high tides, and even nor'easters. All of these occasions, along with long-term and habitat specific sampling, allow us to appreciate the complex dynamics of this relatively unaltered system. At the same time, this location allows us to test different approaches that enhance our understanding: through analysis of a fish's history by examining its earbones, acoustic tracking of fishes and horseshoe crabs, frequent application of sidescan sonar, aerial imagery, and fiber optic cables connected to the adjacent ocean. More importantly we can grasp how the system is changing with the warming climate and rapidly increasing sea level rise. We don't need to know how things are changing in the Arctic and Antarctic; we can see change happening in our backyard.

Throughout my life and career I have always wanted to convey what I have learned about the natural environment and particularly the unique places where fresh and salt water meet in estuaries. This was reflected in my teaching at all levels from middle school into high school, eventually as a graduate student, and most clearly as a university professor. In this position I had the opportunity to share my thoughts and ideas with undergraduates, summer interns, graduate students, postdoctoral students and most clearly with my colleagues at the university and elsewhere in the world.

One prominent way I have shared these thoughts is through the written word. Initially, this was through the scientific literature including several books mainly focused on estuarine fishes along the East Coast of the US and, most recently, the world. In recent years, I have tended to focus on articles and books that attempt to communicate to the average reader the history of research at the Marine Field Station, the ecology of a special estuary—the Mullica River–Great Bay Estuary—and the view of coastal southern New Jersey from the air. In this book, I focus on what we have learned in over three decades on the marshes and waterways bordering Seven Bridges Road and attempt a look into the future. In all these instances, I rely on data, observations, images, and prior publications from the Sheepshead Meadows. When I rely on publications from other areas, I provide citations. All of these are provided by chapter at the end of the book. The scale and scope of this study is augmented by three appendices. These act as bookends for the work, prepared by other authors, addressing details of Lenni-Lenape settlements and development of advanced oceanographic techniques along Seven Bridge Road. A third appendix provides insights into the area through the eyes of a fisherman.

Kenneth W. Able

Chapter 1
Introduction

The Sheepshead Meadows and adjacent islands (Seven Islands/Story Island complex) comprise the 5,982 acres of the Great Bay Wildlife Management Area (Fig. 1.1, 1.2). They share meadows and islands, and many of the characteristics of coastal New Jersey, but they differ in other important ways, including low human population density and low human impacts. They also share numerous protected marshes and waterways that are recognized for their uniqueness, not only for the coast of New Jersey, but the entire East Coast of the U.S. as well. As a result, these meadows and islands provide an exceptional baseline for understanding the geology, structure, and function of New Jersey's coastal landscapes, both back in time and into the future. This baseline is natural history. This approach was the essence for the beginning of "ecology" as we know it today, but the time for natural history studies is often considered past. We are realizing that it needs to be resurrected. Its essence is based on the following questions.

- What plant or animal is it?
- Where does it live?
- How many are there?
- How does it survive and reproduce?
- How did it come to be like it is and live where it does?

The questions above are answerable because these meadows and islands have served as focal points for extensive research, education and outreach by a number of universities (Rutgers University, Stockton University, University of Maryland, Georgetown University, University of Denver, St. Joseph's University) and state, federal, and private institutions (New Jersey Department of Environmental Protection, Barnegat Bay Partnership, Jacques Cousteau National Estuarine Research Reserve, Tuckerton Seaport). As an index of this activity, these same meadows and islands, and

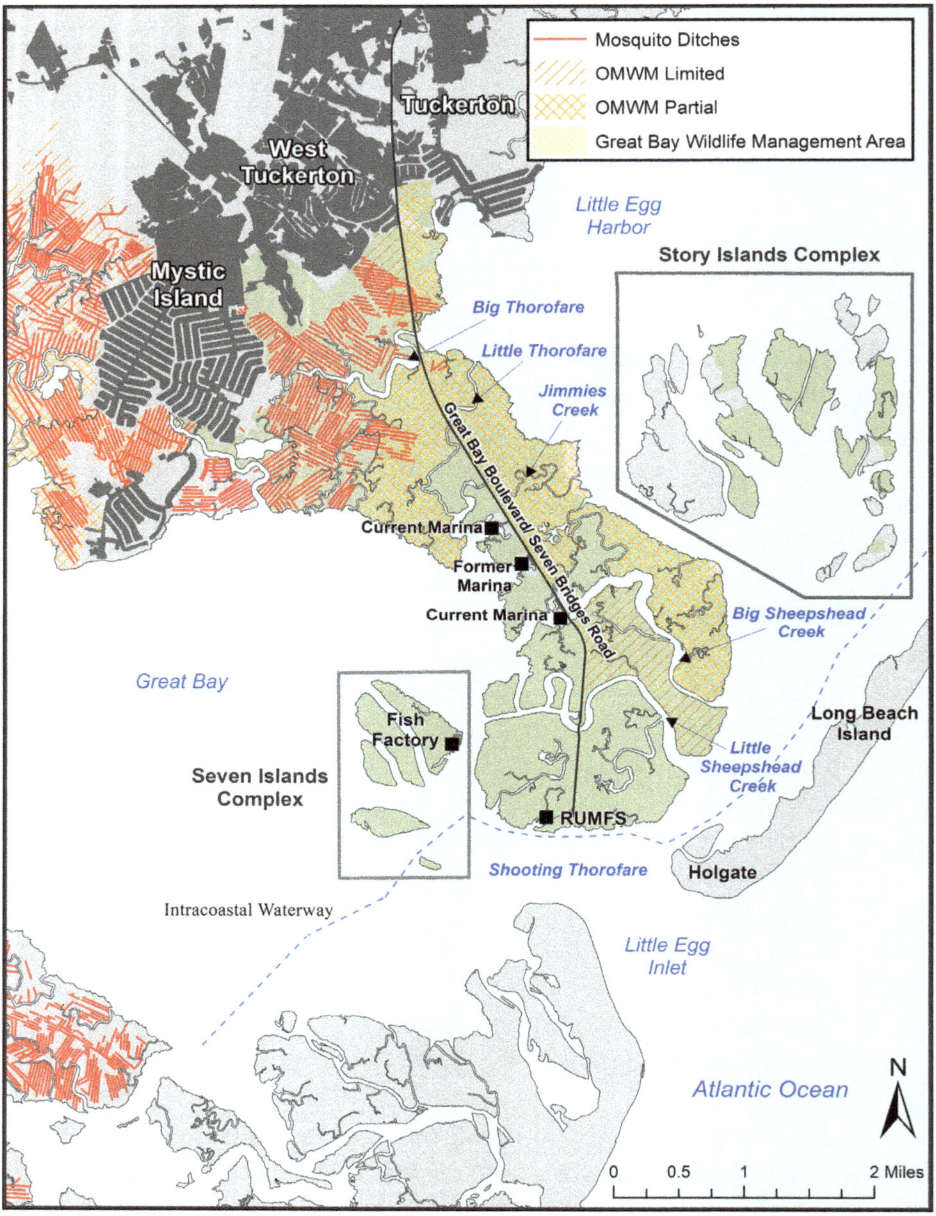

Figure 1.1 Details of the Sheepshead Meadows between Great Bay, Little Egg Harbor and the Atlantic Ocean. The Meadows dominate the area along Seven Bridges Road (officially Great Bay Boulevard) and include human alterations (mosquito ditches, Open Marsh Water Management [OMWM]) all within the Great Bay Wildlife Management Area.

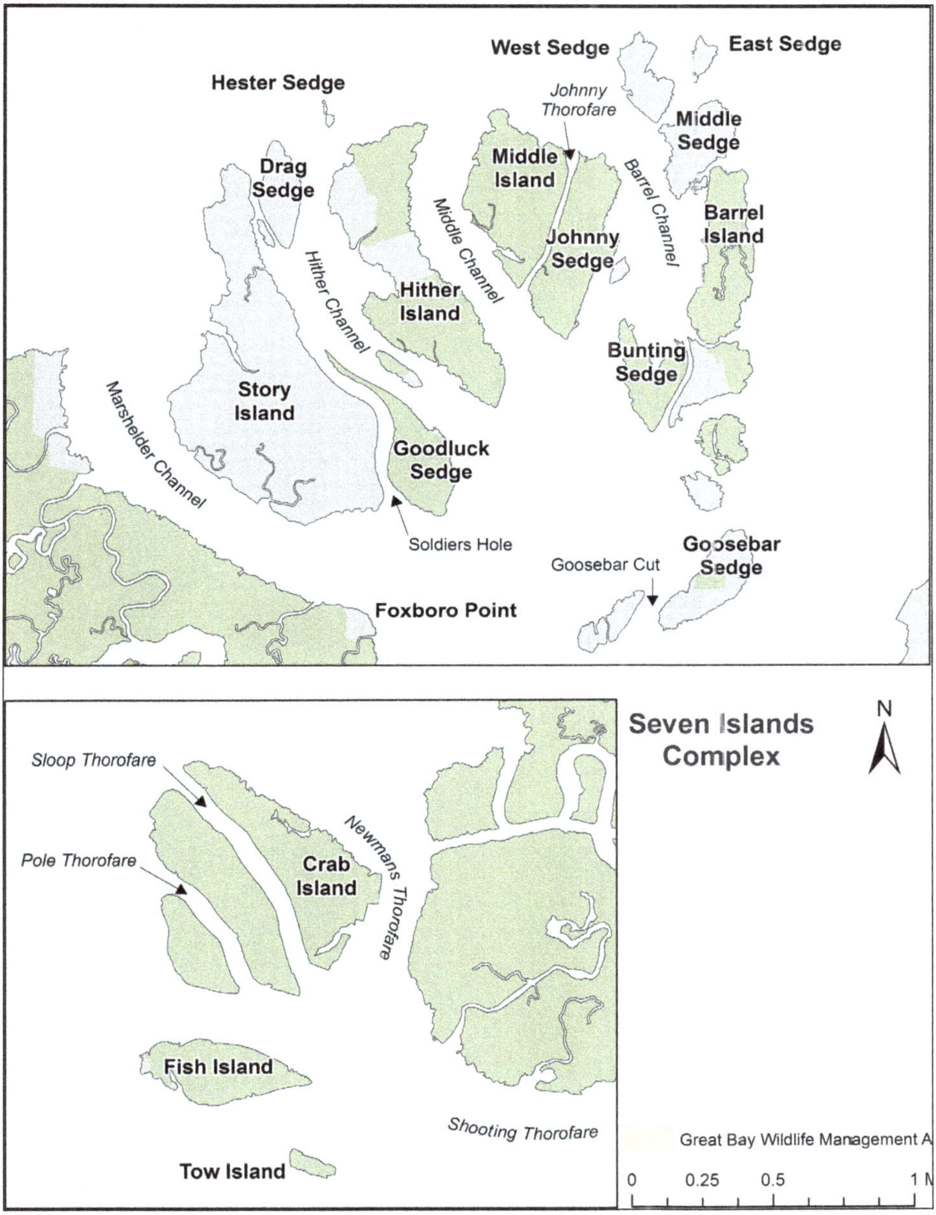

Figure 1.2 Details of the island complexes (Seven Islands, Story Island) adjacent to the Sheepshead Meadows including those portions in the Great Bay Wildlife Management Area.

their importance, are indicated "By the Numbers" at the Rutgers University Marine Field Station (Fig. 1.3). All of these activities and people have provided an exceptional understanding of the human and natural history in this part of the New Jersey shore.

Figure 1.3 Selected research and outreach activities and publications by the Rutgers University Marine Field Station (RUMFS) By the Numbers.

This account is built upon the relationship to prior treatments including those focused on the history of Station 119, now the Rutgers University Marine Field Station, the natural history along the adjoining estuarine waters, and earlier accounts focusing on the fish of the region (see Chapter References). The purpose here is to expand upon these accounts and provide an even broader perspective of its past and extend this perspective into the future. No account of the Jersey shore, however, is complete without an incorporation of human influence. These include the earliest inhabitants, the Lenape Indians and their ancestors, through the time of European settlement and on into the future with a changing climate.

Such an accounting must include the marsh and island habitats, and the diverse microhabitats that comprise the whole of this diverse and relatively unaltered system with its ranging salinities, temperatures, and the marked transition between seasons. All these, in turn, resonate with the especially dominant marshes (Fig. 1.4), and the numerous waterways that surround the area and flow under Seven Bridges Road. In turn, these serve to connect this area with the broader ecosystems of adjoining Great Bay, Barnegat Bay, and the Atlantic Ocean, through Little Egg Inlet, and the Pinelands upstream via the Mullica and Wading Rivers. The annual and seasonally varying fauna reflects these connections in the water, on the land, and in the

Figure 1.4 Aerial view of the Rutgers University Marine Field Station surrounded by unaltered marshes and clean waterways at the southeastern end of the Sheepshead Meadows with Little Egg Inlet and the Atlantic Ocean in the background. Drone image by Thomas Johnson.

air. This is especially evident by the seasonal movements and migrations of the birds, fish, crabs, insects, as well as the mammals that reproduce, grow, and survive in these naturally stressful but productive habitats.

All of these, however, are transitioning as the world's climate changes and sea levels rise. This is reflected in the "low country" along Seven Bridges Road (Fig. 1.1, 1.2) and will continue to be reflected in the meadows and waterways touching and passing under this road. All of these natural and human influenced sources of variation (historically and today) in a relatively undisturbed portion of the coast provide an exceptional opportunity to combine past, present, and future insights into an improved understanding of coastal ecosystems. Scientists would call this an opportunity to understand the "Ecology of Place," i.e., to pursue a general understanding through the detailed understanding of a particular place. This approach incorporates a spatial location and time (frequently long-term). This place-based approach is best practiced at field stations, as at the Rutgers University Marine Field Station (Fig. 1.5). The numerous and diverse research sites and activities in the Sheepshead Meadows and the adjacent islands and waterways near the Station (Fig. 1.6, 1.7) serve to illustrate this approach. In the process, these

Figure 1.5 A view of the Rutgers University Marine Field Station from the surface of nearby Hatfield Creek.

meadows provide a place for teaching and transferring our understanding to coastal communities in general.

In many ways the breadth of exposure to Seven Bridges Road in South Jersey is reflected in the Stephen Young song, "Seven Bridges Road":

> There are stars
> In the Southern sky
> Southward as you go
> There is moonlight
> And moss in the trees
> Down the Seven Bridges Road
>
> Now I have loved you like a baby
> Like some lonesome child
> And I have loved you in a tame way
> And I have loved you wild
>
> Sometimes there's a part of me
> Has to turn from here and go
> Running like a child from these warm stars
> Down the Seven Bridges Road
>
> There are stars in the Southern sky
> And if you ever decide
> You should go
> There is a taste of thyme sweetened honey
> Down the Seven Bridges Road

I enjoyed this song, and especially the Eagles version of it, even before I knew that a road of the same name existed in South Jersey. Numerous diverse groups and individuals, from Joan Baez, King Calaway with Ricky Skaggs, Dolly Parton, Keith Urban and Sugarland, and Alan Jackson to name a few, have covered this song. The opening lyrics are particularly appropriate because the stars in the sky over southern New Jersey are often evident during weekly sampling at night for larval fish over the last 30-plus years from the last bridge on Seven Bridges Road over Little Sheepshead Creek. There are other vantage points or "places" in this low country, however, for both day and night observations, including the other bridges. The setting of Seven Bridges Road has inspired other songs, such as George Wirth's "The Lights of Brigantine," and Ben Sollee's "Slackwater," and likely will inspire others in the future.

Despite the familiarity of "Seven Bridges Road" to many from the area, there are only five bridges over creeks on the Sheepshead Meadows and that, apparently, is all the spans that ever existed on the road (Fig. 1.6). The New Jersey State Highway Department developed plans for two other bridges intended to connect the mainland from Tuckerton to the barrier islands at Beach Haven on Long Beach Island, and Brigantine on the way to Atlantic City.

Surveyors blazed and workmen began constructing Seven Bridges Road across Sheepshead Meadows until it reached the water's edge at Shooting Thorofare in Great Bay in 1929. The project included building five different

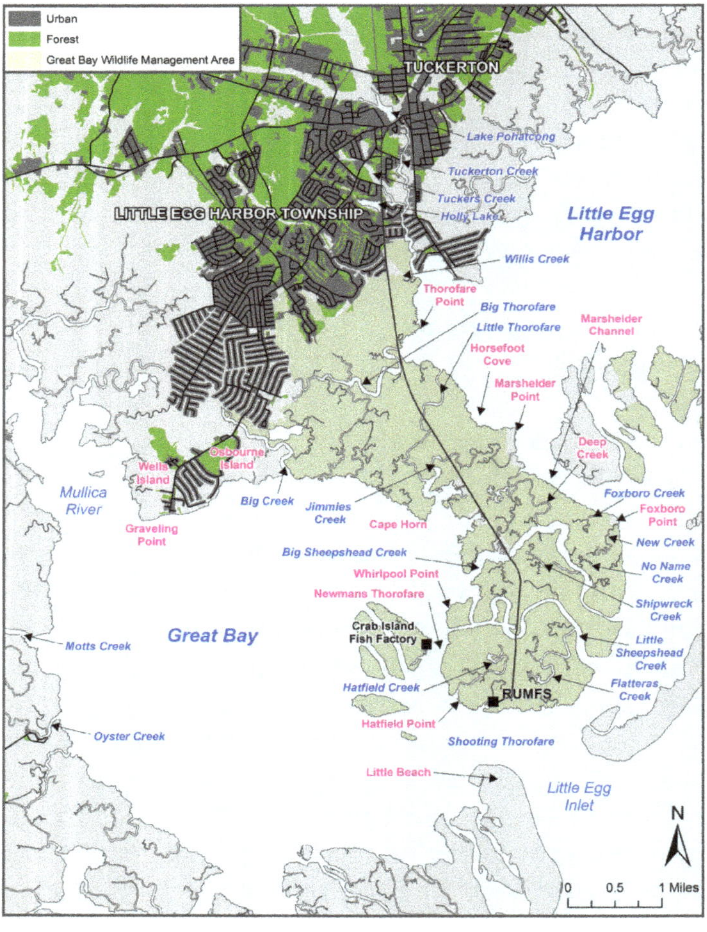

Figure 1.6 Small scale features and place names in the Sheepshead Meadows and adjacent islands in the Great Bay Wildlife Management Area. See Fig. 1.2 for names of islands.

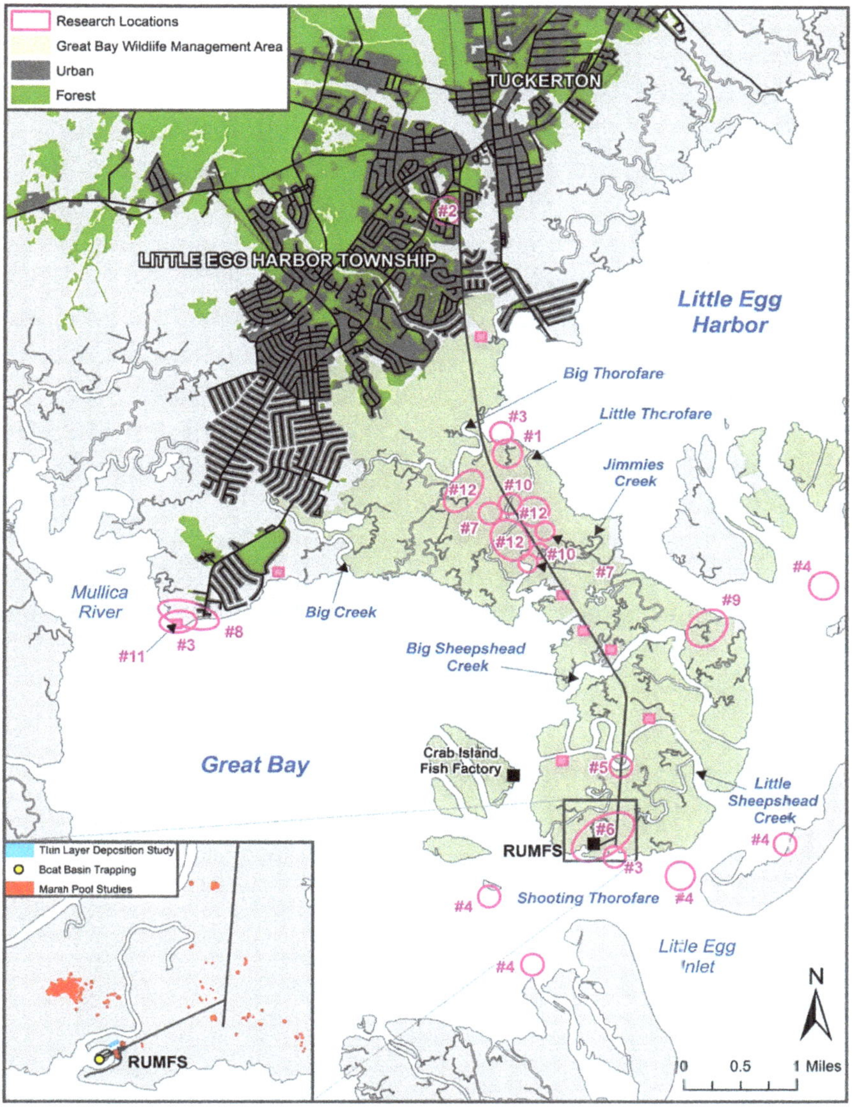

Figure 1.7 Research locations in the Sheepshead Meadows area during the last four decades including 1) effects of ditching on marsh ecosystems, 2) Grassle Pocket Marsh and Jacques Cousteau National Estuarine Research Reserve meteorological platform, 3) RUMFS regular seine sites, 4) other RUMFS seine sites, 5) long-term larval fish sampling site, 6) RUMFS location research site (see inset), 7) OMWM evaluation sites, 8) early oyster studies by Rutgers University, 9) block net and gill net sampling sites, 10) alternative larval fish sampling sites, 11) JC NERR vegetation transects and sedimentation sites (filled squares), and 12) marsh insect and spider community study sites.

bridges spanning the various watercourses that crisscrossed the sedge islands. Despite the state not completing the route to Brigantine, the extant boulevard provided easy access to the United States Coast Guard when it established its new Station 119 at the end of the roadway in 1936. The federal government purchased the station site from the Joseph Wharton estate in January of that year.

At the termination of Seven Bridges Road is the old Coast Guard Station, which needed this advantage in visibility for their life-saving duties. The Station, with its prominent cupola, occupies the highest point in this low country. Rutgers University acquired the former Station 119 in 1972, restored the facility with its multiple windows, and established the Rutgers University Marine Field Station (Fig. 1.5). Appreciation of the location is complemented by traveling along Seven Bridges Road at multiple speeds (i.e., automobile, bike, walking on land, and boating, kayaking, wading) and by aerial views from helicopters and drones. These views are further augmented and interpreted from a historical perspective. All of these approaches, at these diverse locations, will be shared in subsequent chapters.

Chapter 2
Geological and Early Human History

Geological Setting

The recent geological history of the Sheepshead Meadows and the adjacent islands bear witness to the dramatic changes that have occurred during and after the last ice age and up to the present (Fig. 2.9, at end of chapter). Glaciers reached northern New Jersey in three different ice ages that occurred between 80,000 to 11,000 years ago at a time when periglacial (cold but non-glacial) conditions dominated the upper Mullica River Valley. Sea level fell during this period of ice accumulation such that the New Jersey shoreline was as much as 80–100 miles east of its current location. When the ice melted after the last glacier, the rising ocean slowly created the barrier islands that we know today as Long Beach Island and Little Beach. At the same time, the rising seas drowned river valleys such as the Mullica Valley.

Over the last 6,000 years or so, sediments accumulated in the lagoons behind the barrier islands. Forests formed along the freshwaters of the paleo-Mullica River circa 4,000 years ago. Work crews discovered one piece of evidence for the forests during the Garden State Parkway's construction over the Mullica River. At that time, ancient Atlantic white cedar, which can only grow in freshwater bogs, were dredged up from a depth of 40 feet in the river channel (Fig. 2.1). Thus, their occurrence at that location implies that the river and its tributaries were freshwater while today, the salinities are much higher. This pattern of ancient Atlantic white cedar is consistent with observations in the last decade that demonstrated the presence throughout much of the present-day Mullica River and its tributaries, as well as in the adjacent Tuckerton Creek.

The base of the Sheepshead Meadows peninsula was forming at a similar time, about 5,000 years ago. This has been revealed through a series of long sediment cores sampled through the meadows. At the base of these cores is a thin layer of sand at a depth of about 40 feet. A freshwater peat layer tops the sand. Together, this stratigraphy suggests that the peninsula existed as

AGES OLD—Dredging for Garden State Parkway uncovered these trunks of stately cedar trees, remains of prehistoric forest, as much as 40 feet below surface of Mullica River marshes.

Forest 40 Feet Under Ground

Figure 2.1 Newspaper article from February 26, 1954, *Newark Evening News*, documents the dredging up of rot-resistant Atlantic white cedar trees from 40 feet under the marsh during the construction of the Garden State Parkway bridge over the Mullica River just upstream of the Sheepshead Meadows.

a low upland between the paleo-Mullica River and present-day Tuckerton Creek-Little Egg Harbor. As sea level rose after the glaciers retreated, the rising waters flooded the peninsula and allowed salt marshes to form. The depth of the sediments under the Sheepshead Meadows is evident from underwater multibeam images that clearly show the layers of sediment that were deposited to form the meadows (Fig. 2.2). Further, we know that the sediments reach a depth of 30 feet because a recent analysis found that the pilings on which the Rutgers University Marine Field Station sits are at least that long before they reach the sandy substrate that supports them. These salt marshes have kept up with sea level rise until a recent buildup of greenhouse gases in the atmosphere has caused a dramatic increase in sea level, resulting in the marshes that exist today (see Chapter 10).

Over the last several centuries, the inlets providing tidal exchange between the Sheepshead Meadows area and the ocean have also changed dramatically (Fig. 2.3). One of the earliest maps, created by the British in 1769, shows only two inlets: one to the south of Little Beach (now Brigantine or Wreck Inlet) and the other through Long Beach Island. Rather ephemeral in nature, the latter inlet closed and then opened again before eventually closing completely in 1874. Subsequently, it would be referred to as the Old Inlet and then later, when it reopened, as Beach Haven Inlet. Between these was an island known, at the time, as "Mikannow Shoal." In

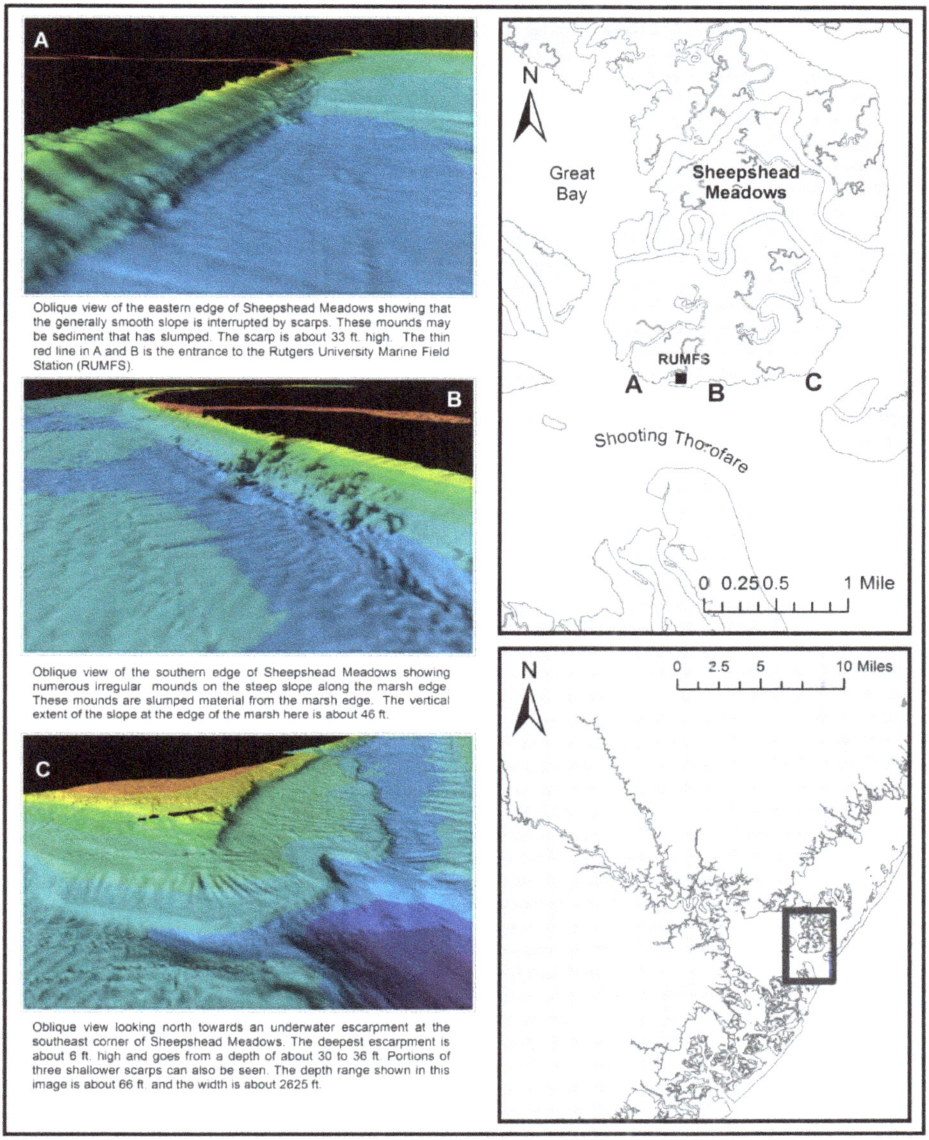

Figure 2.2 Underwater topography (A, B, C) in the region where the Sheepshead Meadows meet the Shooting Thorofare near Little Egg Inlet at the entrance to Great Bay (see insert). Images generated by multibeam sea-bed mapping techniques. From Able (2020).

1800, a channel called New Inlet cut through this island/shoal. It may have formed because Old Inlet was closing. New Inlet subsequently became the major inlet in the region and this continues to the present; it is now known

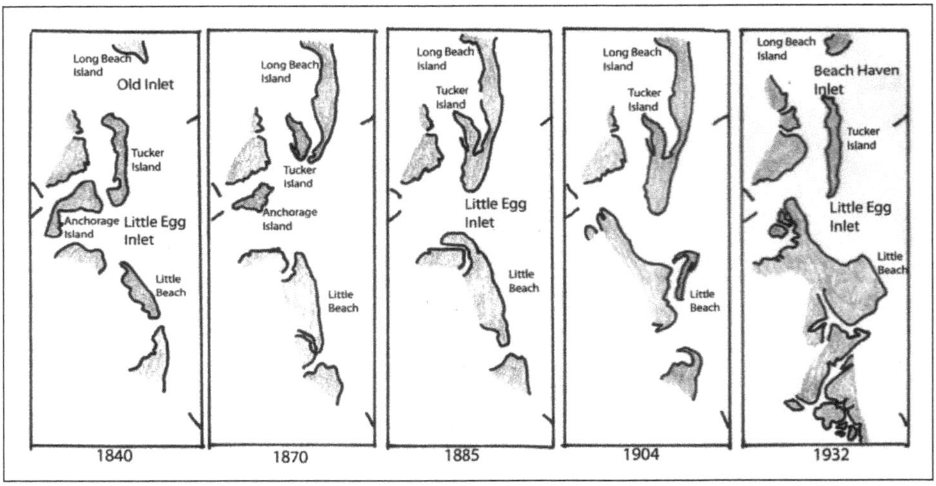

Figure 2.3 These maps show coastal changes in the Long Beach Island-Little Beach area between 1840 and 1932. Blackman (1963) suggests that "Old Inlet" was present as early as 1614. Old Inlet disappeared, then later reappeared, at which point it was referred to as Beach Haven Inlet.

as Little Egg Inlet. Over time, several thorofares were carved through the adjacent Sheepshead Meadows. These changes influenced the early history of humans in the region, and subsequently during the Revolutionary War.

The islands to the north and south of the Sheepshead Meadows peninsula (Fig. 1.2) formed as a flood tidal delta from sediments the ocean carried into the estuary at newly formed inlets. The Story Island complex in present day Little Egg Harbor may have developed as Beach Haven Inlet (or Old Inlet) was formed. The Seven Islands in Great Bay probably formed via similar processes. This pattern of flood-tidal deltas behind inlets, both past and present, is a common phenomenon elsewhere in coastal New Jersey.

Early Humans and the Tuckerton Shell Mound

It was in this geological setting that the earliest humans occupied New Jersey and the Sheepshead Meadows. The source of these people and the timing of their occupation, however, is not completely resolved. Current estimates of humans entering the New World are from 12,000 to 40,000 years ago, but with each new discovery the entrance gets moved further back in time. In addition, many of these estimates for New Jersey may be confounded because the earliest sites on the East Coast of the United States may have occurred when sea level was much lower. Now, many of these sites are likely underwater on the adjacent continental shelf.

The appearance of New Jersey's indigenous peoples occurred during several epochs. These epochs are typically identified as Paleo Indian, Archaic, and Woodland. These labels can be applied to archaeological studies in the Sheepshead Meadows. The first natives to reach New Jersey may have done so before or during the Paleo Indian period when sea level was rising and the post-glacial environment effected radical changes along the New Jersey shore. During the Archaic Period (8,000–1,000 B.C.), the indigenous peoples consisted of small groups of hunters, fishers, and gatherers which may have preferentially settled, either seasonally or permanently, on high ground overlooking coastal estuaries. In the following Early and Middle Woodland Period (3,000 B.C.–1000 A.D.) settlements were larger and the inhabitants less mobile. Shellfish provided an important food source, along with small animals and plants. This period saw the development of smaller projectile points for hunting and fishing. Other studies suggest that the bow and arrow emerged about 1000 years ago with even smaller projectile points. It was during the Middle Woodland period that the Tuckerton Mound began to be regularly used. The Woodland Period ended with the Contact Period when the first European settlers arrived.

This background sets the archaeological stage for the extensive studies of the natives in the Sheepshead Meadows area. One of the best preserved and most frequently studied Native American sites in New Jersey, the Tuckerton Shell Mound (Fig. 2.4), first gained scientific recognition in 1868. Located on the edge of Tuckerton Cove in Little Egg Harbor Township, Ocean County, it remains extant about 100 feet from Seven Bridges Road (Fig. 2.4). Subsequent studies (Appendix 1) have confirmed that it served as a fishing station with the cultural remnants consisting mostly of hard clam shells, but other species, in small quantities, are also present (conch, oysters, bay scallops). Many other shell mounds existed around Great Bay and along the coastal zone, later providing convenient sources of lime for flux in iron furnaces such as those at Batsto, resulting in many shell mounds/middens being destroyed. Radiocarbon dating from clamshells at the Tuckerton Mound, sampled in 1974 at 8 feet below the surrounding salt marsh, indicates an age of approximately 1530 years before present (B.P.). Subsequent investigations that the Archaeological Society of New Jersey conducted yielded charcoal from the mound's base with an age of approximately 1900 years. The same study demonstrated that the shell

Figure 2.4 Drone aerial image of the Tuckerton Shell Mound on January 15, 2018. This site is located about 100 feet from Seven Bridges Road with trees and power poles lining the road (upper left). At this time the mound is covered with one prominent dead red cedar, living red cedars and deciduous trees lacking leaves at this time of the year. The linear white lines around the mound are mosquito ditches filled with ice and covered with snow. Drone image by Thomas Johnson.

mound's base rests on firm mineral sediments as much as 14 feet below the meadow's surface. The essence of previous studies found that natives used the site for shellfish harvesting and perhaps processing/smoking the clam meat. Sediment accumulation during the mound's accretion indicates that sea level has risen markedly during the last 2000 years of its existence.

Further studies in the immediate vicinity near Seven Bridges Road support this interpretation and provide an improved understanding of the activities of those who occupied the area, including the Lenape in the Late Woodland period. It appears the indigenous peoples used the Tuckerton Mound site during multiple seasons and, perhaps, longer-term by the occupants of several nearby sites. The most extensively studied area near the Tuckerton Mound, the Pennella Site (Fig. 2.5), suggests occupation at different times, ranging from the Late Paleo Indian, Late Archaic, to the Late Woodland periods, with the main site occupation occurring during the Middle Woodland, based on the most recent analysis (see Appendix 1). The most extensive use, based on radiocarbon dating, ranged from 1530–1810 B.P., thus with overlapping dates from the Tuckerton Mound, indicating concomitant usage for these sites.

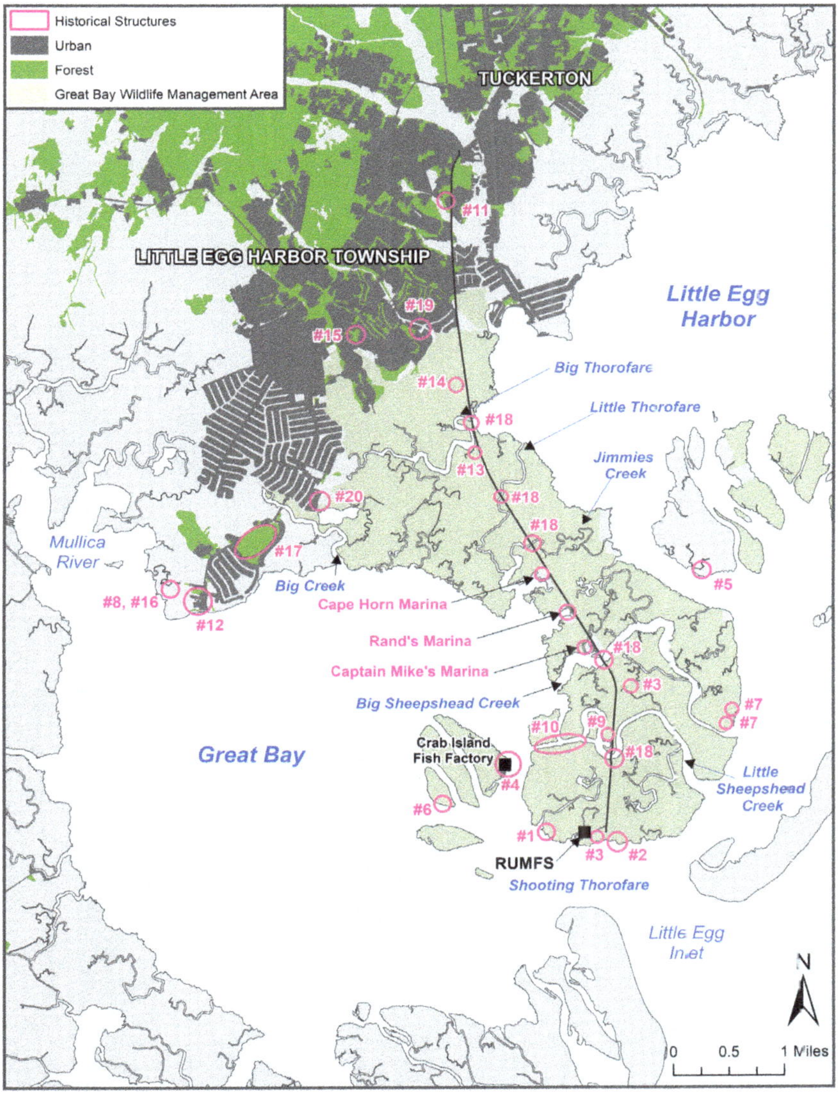

Figure 2.5 Locations of historical structures in the Sheepshead Meadows and the adjacent islands. 1) Hatfield Hotel, 2) Willit Hotel, 3) square marsh pool, 4) Crab Island Fish Factory, 5) Story Island Fish Factory, 6) Little Crab Island Fish Factory, 7) Sheepshead North and South Fish Factories, 8) Wells Island Fish Factory, 9) Fish Factory Weigh Station, 10) Fish Factory dredged Channel, 11) RUMFS Annex/Jacques Cousteau National Estuarine Research Reserve, 12) Hotel Bivalve (approximate), 13) former First Bridge Marina, 14) Tuckerton Mound, 15) Pennella Site, 16) Wells Island, 17) Osborne Island, 18) bridges, 19) Willit/Andrews farm house and wireless tower.

During this time period, the Pennella Site, which lies approximately 3000 feet to the northwest of the Tuckerton Mound on the first high elevation (about 30 feet above mean sea level) (Fig. 2.5), served as a seasonal base camp and possibly also as a year-round camp. This interpretation is supported by a human burial area and evidence of different types of hunting (Fig. 2.6) and gathering activities, including pieces of pottery, with some sign of agricultural activities including hearths, food storage pits, and a possible subterranean house. Speculation suggests that a group of families of up to twenty members occupied the Pennella site during the Middle Woodland Period. These numbers may have increased during the summer, when they were gathering hard clams and processing them at the Tuckerton Mound site. At the Pennella Site, however, oysters comprise the most abundant shellfish remains. Other shellfish there included whelks, ribbed mussels, slipper shells, and hard clams. The Tuckerton Historical Society displays many of these artifacts, derived from the archeological work of Andrew Stanzeski (see Appendix 1).

In addition to shellfish, diamondback terrapins and their eggs could have been easily gathered. Other food items included box turtles (Fig.

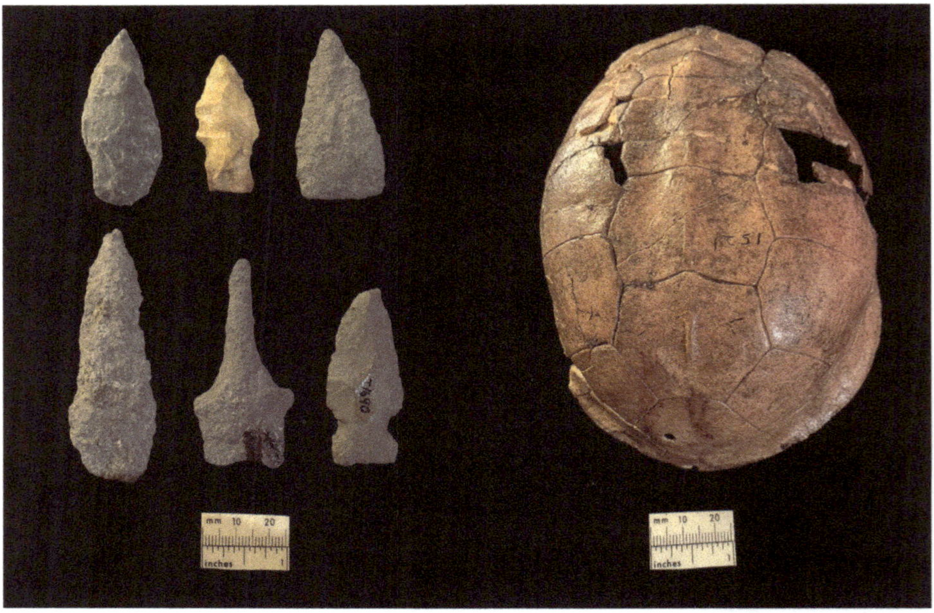

Figure 2.6 (left) Arrow points retrieved at the Pennella Site. Figure 2.7 (right) Box Turtle shell retrieved at the Pennella Site. Photos from Andrew Stanzeski.

2.7), spotted turtles, and snapping turtles. Evidence of other hunted species included white tailed deer, black bear, ducks and geese, and possibly wolves. Gathering of plants, mushrooms, seeds, and nuts also occurred, based on evidence at the site. Fish remains at the site have been found, but their relative fragility may result in underestimating their importance. Present species recorded from there include black drum and sheepshead. The likelihood that fishing provided an important way to gather food is based on the discovery of a dugout canoe (Fig. 2.8) at nearby Beach Haven, on the ocean side, dating from 1810 to 1530 B.P. and thus in the temporal range for occupation at the Tuckerton Mound and the Pennella sites. The Atlantic white cedar canoe was constructed by chopping, burning, and scraping. The annular rings in the wood indicated that the tree was at least 115 years old when felled.

Other evidence of Lenape occupation includes the main encampments at Wells Island and Osborne Island, which are located directly adjacent to each other and within a mile of the Mound (Fig. 2.5). They also had a number

Figure 2.8 Remains of dugout canoe discovered in the ocean off Beach Haven. Now in the Beach Haven Museum, as shared by Ron Marr.

of discreet shell mounds primarily dominated by oyster shells, but hard clams were also present as well as mussel shells, barnacles, and unidentified bones. Other faunal remains from the site included small mammals (shrew, river otter, vole, skunk), birds (goose, turkey), reptiles (spotted turtle, box turtle, snake), amphibians (toads), as well as some fish. Radiocarbon dates from both sites are similar to the Late Woodland Period (Appendix 1). An Osborne Island burial site ranged from approximately 400–720 years ago. The Wells Island site had a similar date of approximately 650 years.

It is not clear whether the local Lenape occupants used the Pennella, Wells and Osborne island sites concomitantly during the Contact Period, but this appeared to have occurred just to the north at the West Creek Site on Barnegat Bay. In general, European diseases and the loss of land to European colonists devastated the region's Lenape population, and they departed from the area, migrating to such reservations as Brotherton and to points further west of New Jersey. During their tenure, the Willit/Andrews family cleared and plowed the Pennella Site, but the site has not been used for farming since the 1930s. During the mid-1970s until 1989, a large development occupied the land, built on the site as part of the Mystic Islands construction, but analysis of the Lenape relics and early European settlers continues (Appendix 1).

Figure 2.9

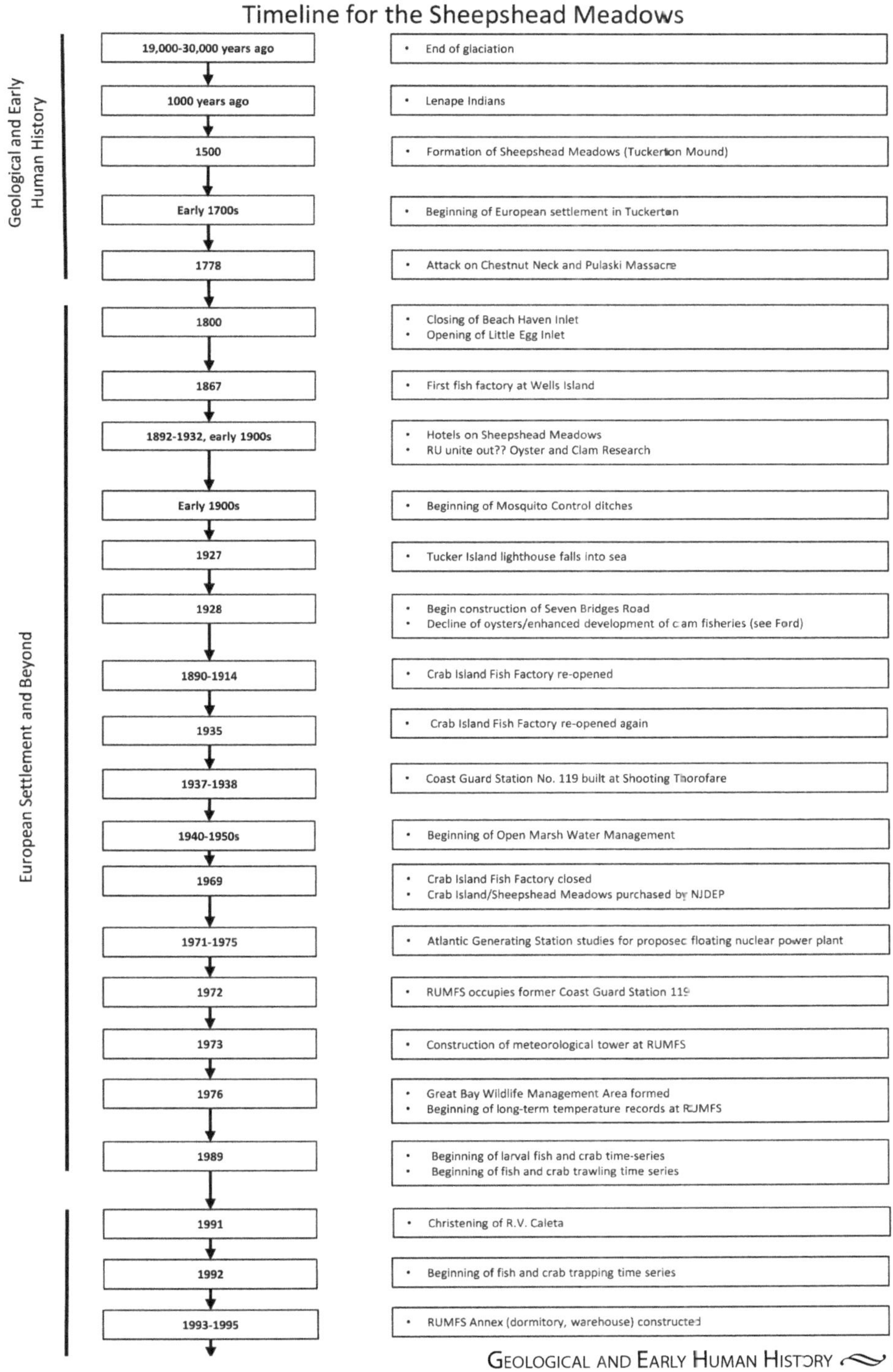
Timeline for the Sheepshead Meadows
Geological and Early Human History
19,000-30,000 years ago
End of glaciation
1000 years ago
Lenape Indians
1500
Formation of Sheepshead Meadows (Tuckerton Mound)
Early 1700s
Beginning of European settlement in Tuckerton
1778
Attack on Chestnut Neck and Pulaski Massacre
European Settlement and Beyond
1800
Closing of Beach Haven Inlet
Opening of Little Egg Inlet
1867
First fish factory at Wells Island
1892-1932, early 1900s
Hotels on Sheepshead Meadows
RU unite out?? Oyster and Clam Research
Early 1900s
Beginning of Mosquito Control ditches
1927
Tucker Island lighthouse falls into sea
1928
Begin construction of Seven Bridges Road
Decline of oysters/enhanced development of clam fisheries (see Ford)
1890-1914
Crab Island Fish Factory re-opened
1935
Crab Island Fish Factory re-opened again
1937-1938
Coast Guard Station No. 119 built at Shooting Thorofare
1940-1950s
Beginning of Open Marsh Water Management
1969
Crab Island Fish Factory closed
Crab Island/Sheepshead Meadows purchased by NJDEP
1971-1975
Atlantic Generating Station studies for proposed floating nuclear power plant
1972
RUMFS occupies former Coast Guard Station 119
1973
Construction of meteorological tower at RUMFS
1976
Great Bay Wildlife Management Area formed
Beginning of long-term temperature records at RUMFS
1989
Beginning of larval fish and crab time-series
Beginning of fish and crab trawling time series
1991
Christening of R.V. Caleta
1992
Beginning of fish and crab trapping time series
1993-1995
RUMFS Annex (dormitory, warehouse) constructed

The Modern Coast

1996	• Beginning of fish and crab trawling time series
1996-2011	• Fiber optic link to LEO-15 installed and operated
1996	• Christening of R.V. Arabella at RUMFS
1997	• JCNERR formed • Striper tracker operation from xxxx-xxxx
1998	• First Year in Life of Estuarine Fishes book
2007	• Removal of meteorological tower at RUMFS
2012	• "Hurricane" Sandy
2015	• Publication of Station 119: A History of RUMFS
2020	• Publication of Beneath the Surface book
2022	• Publication of Estuarine Fish and Fisheries book

Chapter 3
European Settlement through the Revolutionary War and Beyond

Settlement Along the Shore

For the purposes of this chapter, the history of Seven Bridges Road, the surrounding islands, and adjacent areas, experienced events that also occurred in the broader region. These events begin with European settlement and continue through the Revolutionary War and beyond, but before the road's completion (see Fig. 2.1). The European history begins in 1609 when the Dutch vessel, the HALF MOON, sailed by the inlets in what is now South Jersey. Long boats from the ship rowed to the shore for food, resulting in the name "Eyren Haven," the Harbor of Eggs, due to the large number of nesting birds that yielded eggs for the sailors. These early sailors likely accessed the area through Old Inlet (Fig. 2.1) because Little Egg Inlet did not exist at the time. Its formation occurred around 1800. The bird eggs' importance also provided the impetus for the early name of the Mullica River and Great Bay as the Little Egg Harbor River. Of related note is the early use of the term Egg Harbor, an area that now includes Little Egg Inlet, Shooting Thorofare, and Great Bay, as well as the southern end of Barnegat Bay or Little Egg Harbor. Further south is Great Egg Harbor and the Great Egg Harbor River. The Sheepshead Meadows lies in the approximate center of these waters, surrounded by water on three sides. The waters around these meadows served as major conduits for shipping, other forms of commerce, and military activities. Numerous events of historical significance occurred within sight of these meadows. In addition, events in nearby Tuckerton, or its older moniker Clamtown, had a major influence on the meadows.

Understanding the relationship between the Sheepshead Meadows and the surrounding waterways begins with comprehending how the earliest Europeans settled and harvested the region's natural resources. Settlement in Tuckerton occurred when members of the Religious Society of Friends, or Quakers, arrived as early as 1700. The settlers likely initially continued many of the same activities as the Lenape, i.e., fishing and hunting for food in the meadows and the waterways around them, including the adjacent ocean for whales. The Tuckerton Mound and similar deposits on nearby Osbourne and Wells islands provide visual evidence of indigenous shell-fishing activities. Continuing this tradition, the early European colonists probably continued gathering/harvesting intertidal oysters and hard clams by "treading" for them, an easy and rewarding activity. Other benefits from the adjacent waters included Atlantic menhaden or mossbunkers, which were likely caught with nets or weirs and used as food, and later as fertilizer as agriculture developed. It is likely that they also used the meadows for grazing livestock, a well-documented practice in the nearby Pleasantville Meadows at about this same time.

The Europeans brought a more extensive agriculture tradition to the area during the earliest stages of settlement. The New Jersey Historic Preservation Office developed a fairly detailed account of the Willits/Andrews farm on Osbourn Island at the fringes of the Sheepshead Meadows. Leah Blackman noted that this farmhouse was one of the most significant historic sites representing the period of the area's earliest European settlement. Around 1715, the Willits family migrated from Long Island and settled in the local area. The family remained on the property until around 1831, when the Andrews family acquired the land and maintained it into the late nineteenth century. The farmhouse burned down in the mid-1930s, and most of the area now lies beneath a twentieth-century housing development. During the earliest period, Indian corn and rye dominated crop production, which eventually yielded to wheat and a variety of orchard crops and livestock. The adjacent meadows provided grazing areas and salt hay for the same livestock, especially during the winter. Shellfish and fishing in the nearby waters provided prominent food-gathering activities at the time, with a trace of a road or path from the farm to nearby Tuckerton Cove (Fig. 1.1). Later, during the Andrews' ownership, potato cultivation and dairy and poultry production occurred on the farm. The development of the Willits/Andrews farm was synchronous with the settlement of transplanted Long Island Quakers in Tuckerton, which began about 1700. Shortly thereafter, the new settlers constructed a grist mill and a Meeting House for the worshiping Quakers who dominated the area. Island Road (now Radio Road, see Fig. 1.1) was laid out in 1761 as a result of these developing activities in Tuckerton.

Conflict with the British

Actions by the British during the mid-1700s directly relate to waterway usage around Sheepshead Meadows. Much of this activity stemmed from the desire to tax the colonists for many engagements on the part of the British, including the French and Indian War in the mid-1700s. The Crown made other attempts to pass on the expenses of the British activities to the colonies, including the Sugar Act in 1764 and the Stamp Act in 1765. These highly unpopular actions prompted the colonists to seek financial retribution through smuggling and privateering, occupations initially illegal, but later made legitimate under the justification of "letters of marque" from individual colonies, and later by Congress. These activities occurred in New Jersey and especially in the waters around the Sheepshead Meadows as the area was a major shipbuilding and seafaring region and the local inlets were somewhat treacherous and required extensive local knowledge that the British lacked. Further, once inside the inlet (then the Old Inlet through what is now southern Long Beach Island), the deep waterways, and especially the Mullica River, made navigation possible far inland to locations where the confiscated vessels could be tied up, and their cargo could be offloaded. These sites along the Mullica River included Chestnut Neck, Clark's Landing, Lower Bank, Mordecai Landing, and The Forks. With the cargo removed from the vessels and placed in wagons, teamsters transported the goods through the Pine Barrens to such proximate urban centers as Philadelphia, where the wagonloads sold for a good profit. Some of this cargo reportedly provisioned George Washington's army at Valley Forge.

The privateering activities and transport of the spoils, which often occurred in sight of the Sheepshead Meadows, proved financially draining on the British war effort and their coffers back home, and doubly so due to the numerous British seamen taken prisoner. In the same area, the construction of several salt works in the 1770s, including ones in Tuckerton and Falkinburg Island (now Osborne Island), upstream on the Mullica at Chestnut Neck, and along the Bass River, provided a valuable commodity as salt served as a food preservative. These works provided their owners with a significant profit, especially during the blockade of colonial ports. This period of importance, however, only lasted a scant few years due to salt imports from other countries. By 1778, many saltworks had closed down. These works, including one near Tuckerton, again provided salt during the War of 1812 for many of the same reasons that occurred during the Revolutionary War.

These events and activities during the Revolutionary War provided the impetus for the British attempting to remove the "nest of rebel pirates" operating out of nearby Old Inlet and the Crown's attack on Chestnut Neck further up the Mullica River. In July 1777, a sentinel located on Sheepshead Meadows near present-day Foxboro Point observed an initial British expedition entering Old Inlet. This preliminary British action only triggered an increase in privateering during the ensuing year. In response, a larger British naval expedition arrived off the inlet in October 1778 and proceeded in fits and starts to pass over the treacherous sandbars at the inlet and on up to Chestnut Neck, with groundings of several vessels along the way. Upon reaching Chestnut Neck, the British forces attacked and destroyed the fort and village, and burned and sunk several ships, including some previously captured British vessels. Attempts to refloat the previously grounded ships delayed the retreat back downstream, past the Sheepshead Meadows. In the interim, a British force attacked a group of sleeping colonial soldiers that had been dispatched under the direction of Count Pulaski to nearby Osborne Island, at the site of the Willits Farm, along the fringe of the meadows. These British forces slaughtered the continental troops as they slept, and the British reboarded their ships and retreated out through the inlet. This was the last British military action in the area during the Revolution, and the raid did little to reduce the attacks on British shipping. Once the British general Cornwallis suffered defeat at Yorktown, and the war was coming to a close, the privateering activity drew to a close in the waterways around the Sheepshead Meadows. It resumed, however, during the War of 1812.

Developing American Resources

The more intense development of shellfisheries, which began with the earliest Lenape occupation of the area around Seven Bridges Road, continued in the 1800s, especially for oysters and hard clams (hence, one of the early names for Tuckerton was Clamtown). As a result of this important seafood harvest, Rutgers University researchers became involved in the study and management of these fisheries in the late 1880s, especially oysters. The clearest evidence locally of the Rutgers presence was a number of small buildings about four miles south of Tuckerton in the meadows around Great Bay in the vicinity of Big Creek (Fig 1.1). A nor'easter in February 1920, however, destroyed these facilities. Rutgers carried out research simultaneously on the floating barge CYNTHIA at a variety of locations, including near Tuckerton (on Thompson Creek and Edge Cove) and Turtle Island near Graveling Point at the mouth of the Mullica River, beginning in 1910. Due to overharvesting and destruction of oyster beds in the area, however, the fishery unfortunately declined precipitously by 1930. A similar decline occurred with hard clams.

In the early years, most travel and commerce in the region occurred by boat in the numerous waterways on the outside and through the Sheepshead Meadows. Thus, the lodging for travelers was limited to the edge of the meadows. There is evidence of two "hotels" in the meadows from earlier maps. The Willits and Hatfield hotels stood on the edge of Shooting Thorofare (Fig. 2.5). In addition, a map from 1876 depicts two buildings near Willits Creek, with one located on each side of Tow Island Creek where it empties into Shooting Thorofare. This was at the approximate location of the end of today's Great Bay Boulevard. The locations of these two hotels may presently be underwater, even at low tide, due to erosion all along the Sheepshead Meadows' southern and western margins. Some remnants of the Hatfield Hotel, however, remain evident from the shell pile located on the meadows-side. Today numerous shells, some well-worn bricks and old, broken bottles are all that remain and only visible at low tide (Fig. 3.1). At about the same period, the region's seafaring and shipbuilding traditions appear reflected in the construction of the Bartlett House and the Sea Captain's House (Fig. 3.2). The first is one of the oldest historic buildings on the New Jersey coast. Both are within the Tuckerton Seaport.

Notes from the Road—Beachcombing for History

Archaeological investigations and historical research reveal, in many ways, the history of human existence on and around the Sheepshead Meadows. These include from the presence of indigenous peoples to later detailed written records about pioneer Europeans, such as those of Leah Blackman for Little Egg Harbor Township, and photographs from numerous sources. Another unique source is based on beachcombing at the end of Seven Bridges Road, where sandy wash-over deposits are apparent, and where various forms of "beach glass" have been recovered from the last few centuries. The glass types range from pieces emanating from the 1900s, such as unsurprising screw top bottles, cosmetic containers, Team Cola bottles, and Bennington pottery from fairly recent decades, to "blob top" bottles and distinctive pieces of medicine bottles, along with pieces of salt-glaze pottery from the early 1900s. Other pieces dated to the late 1800s, including yellowware and transferware. The local source for these may have been the two hotels that existed around that time, including the Willits Hotel and Hatfield Hotel at the

Figure 3.1 (left) The remnants from the Hatfield Hotel at the mouth of Hatfield Creek along Newmans Thorofare, with artifacts scattered among numerous shells. Note Rutgers University Marine Field Station in the background.

Figure 3.2 (below) Aerial image of the Bartlett House "Homestead," the oldest house in Ocean County, in the upper left. The large building on a hill in the center of the image is the Sea Captain's House from the early nineteenth century. Both are located in the Tuckerton Seaport. Drone image by Rob Auermuller.

southern end of the Sheepshead Meadows (see Fig. 2.5). One of the oldest artifacts discovered was a clay pipe stem with the maker's mark lettering, "Jean Gambier a Paris," a prolific manufacturer after 1846. Another potential source could have been discards from sailing vessels, both privateers and the British military, from the Revolutionary War era, including the attack on Chestnut Neck up the Mullica River.

As sailing vessels gave way to steam power, the inlets (Old, New, Beach Haven, Little Egg Inlet) continued to be used, particularly by the six fish factories that processed menhaden inside these inlets from the late 1800s up to the 1960s. More recent articles could have originated from Coast Guard Station 119, built in the late 1930s and remaining active into the early 1960s, or possibly the garbage dump on nearby Crab Island that existed from the 1930s into the 1940s.

Local Evidence of the Industrial Revolution

This revolution in the United States began when prior agrarian societies became more industrialized and urbanized. It occurred, generally, from the mid-1760s to the mid-1800s and Sheepshead Meadows reflected these industrial changes to some degree, even though the area was remote and certainly not urbanized. One reflection of that industrialization was the development of Atlantic menhaden processing for a variety of uses. While menhaden were initially often appreciated as human food, when the oil from whaling became scarce, due to overfishing by the time of the Civil War, the oil from menhaden became valuable. This became most obvious in the Sheepshead Meadows with the first factory established for menhaden reduction to create oil and fertilizer at Wells Island near Graveling Point (see Fig. 1.1) in the late 1860s. Two other factories at the mouth of Big Sheepshead Creek near Story Island subsequently developed in the 1870s and yet another at Little Crab Island in the 1880s. All of these were abandoned concomitantly with the construction of the largest factory on Crab Island, around the early 1900s (Fig. 3.3). The process of industrialization is also reflected in the transition from sailing vessels to steam-powered vessels in the early twentieth century (Fig. 3.4). Large numbers of local people and others from more southerly areas manned the factory (Fig. 3.5). In 1912, construction of another factory occurred on Story Island. It persisted until 1938, when a conflagration destroyed the facility and never underwent rebuilding, although the factory chimney stack was used as an iconic landmark for decades until it was destroyed during a storm in April 1954.

Notes from the Road—Menhaden as Human Food

When we think of menhaden or bunker today, we consider them as the oily, easily rotted fish that has been used as an agricultural fertilizer and feedstock, or reduced to number of industrial products, or perhaps as crab bait over the last century. What we forget is that at one time, many appreciated this fish as human food. In New Jersey this occurred most frequently in the early nineteenth century. The small menhaden, after feeding extensively in estuaries like Great Bay and Little Egg Harbor during the summer, were captured, canned, and marketed as "American sardines" and shipped elsewhere in North America and Europe. By fall, the fish are typically large and fat after feeding all summer in estuaries and in preparation for their winter migration to the south. These fish were believed to taste different and as

Figure 3.3 Aerial image of Crab Island from 1998 puts its location into perspective as it is just inside Little Egg Inlet (top) with the Rutgers University Marine Field Station at the end of the Sheepshead Meadows. The abandoned buildings and water tower of the fish factory on Crab Island are in the foreground. An earthern wall or berm surrounds the former garbage dump area. The linear portion from lower left to middle right is the remnant of the former landing strip used by spotter pilots for the menhaden purse seine boats.

fat "like butter" when baked. Others were salted for consumption during the winter. Their value as human food was such that the markets extended from coastal New Jersey, but also Maine, Massachusetts, and New York City, as well as including shipment to the West Indies. Once the menhaden fisheries became industrialized, however, for production of oils, meal, and fertilizer, the dominant way humans continued consuming menhaden was from eating chickens raised on meal made from menhaden as processed in the fish factories around the Sheepshead Meadows.

The variability in the menhaden fishery history also reflects unavailability of menhaden in some years, the closure of the factories during WW I when the federal government confiscated the fishing vessels for the war effort, and the ensuing commitment of the Crab Island Factory location as a trash dump for Atlantic City for several years after WW II. The history of the workers at these plants also reflected the danger these positions entailed with frequent newspaper accounts of deaths by lightning strikes, drownings, and accidents associated with the processing machinery, as well as unexplained mortalities.

Figure 3.4 An example of a steam powered vessel for capturing menhaden for processing located at the Crab Island Fish Factory. Photograph, circa 1906. Courtesy of the Tuckerton Historical Society.

Figure 3.5 Transport of workers to the fish factory on Crab Island circa 1906. Image from the Tuckerton Seaport.

Notes from the Road—Crab Island as a Garbage Dump

Beginning around 1916, Crab Island began serving as a garbage dump. This occurred because of complaints in Atlantic City about the smells that the garbage generated in this growing city. The odor proved less of a problem on Crab Island because the "stink house" at the fish factory there already created its own obnoxious smells. In the new arrangement, scows transported the garbage from Atlantic City through the waterways between Atlantic City and Crab Island. By 1919, newspaper accounts indicate large amounts of garbage had accumulated on Crab Island but not been processed. Perhaps this was the period when construction of an earthen wall or berm occurred around the fish factory and this wall remains evident today (Fig. 3.3). During this time, the accumulated garbage and the associated smells from the fish factory caused nearby Beach Haven on Long Beach Island to initiate a legal attempt to restrain Atlantic City from its continued use of Crab Island as a garbage dump. Beyond the smell, another impact from the garbage dump was the release of grease in large quantities that fouled the waters around Crab Island, making it difficult for clammers to handle their tongs to harvest clams.

At about the same time 60–75 men obtained employment at the garbage dump. Some were migrant workers from more southern areas, but the workforce also included local individuals who had been arrested and were forced to serve their time at the garbage dump on Crab Island. Some of these men had housing in a dormitory above the dining hall. The remnants of the garbage dump can still be found at Crab Island's northwestern portion, where the waters are currently washing away the berm surrounding the former dump. This is evident from the bottles, cans, and other items that can be collected there. The dumping of garbage on Crab Island continued into the early 1930s, but by 1933, the Island Fertilizer Company, which operated the dump, declared bankruptcy. Shortly thereafter, the plans to operate Crab Island solely as a menhaden reduction plant developed and Howard J. Smith leased Crab Island. He began an intensive site industrialization, with new techniques and technologies and so began the building of a mini-city on Crab Island.

The Tuckerton Wireless

Other indications of the "industrialization" in the area around the Sheepshead Meadows (Fig. 2.5) include the development of "wireless telegraphy" and erection of the antenna, the tallest (320 feet) structure in the world at that time, for the Tuckerton Wireless (Fig. 3.6), which provided a level, unobstructed sweep across the Atlantic Ocean. The construction of this structure at Hickory Island, now part of Mystic Islands, began in 1912 and completed in 1914 for the purpose of communication between Germany and the U.S. At that time, the site could only be reached by a corduroy road (called Island Road but changed to Radio Road after the construction of the tower) from Tuckerton. The supports for the tower consisted, in large part, of three concrete, steel reinforced blocks buried 20 feet in the meadows and extended 24 feet above them. During the construction phase, many of the Germans working on the site boarded in Tuckerton and many people from Tuckerton worked at the site. While in operation, the location was a very busy place with 7,360 paid messages transmitted and 6,429 messages received between October 1914 and June 1915.

When the United States entered WW I against Germany, the U.S. government closed and occupied the wireless site. In February 1917, government officials ordered all German employees to leave the site. The site remained in service to the U.S. government for trans-oceanic messages with ships and broadcasting the nightly news. The facility closed down completely in 1949 and underwent demolition in December 1955. The large concrete anchors

Figure 3.6 The Tuckerton Wireless, one of the tallest structures in the world, developed by Germany and France for wireless communication across the Atlantic Ocean. Note the adjoining structures to the left of the base that are still in existence today.

and some of the associated buildings remain extant in the Mystic Islands lagoon development today. In fact, on a clear day, the former chimney on a building near the base of the tower can still be viewed from the first bridge on Seven Bridges Road. A portion of the tower base has been relocated to the Tuckerton Historical Society's front yard.

Ditching for Mosquito Control

In the early 1900s, after the earlier uses of meadows were no longer deemed important, state government often placed an emphasis on "reclaiming" the marshes to make them more economically important. During this period, and extending for several decades, the marshes were modified with parallel grid ditches to help control mosquitoes. The movement of many people to the shore and the growing summer tourism often served as justification for creating these ditches, advocating for the need to reduce the negative impact of mosquitoes. A good example is the extensive ditching that occurred prior to the development of the Mystic Islands at the western limits of the Sheepshead Meadows (Fig. 3.7).

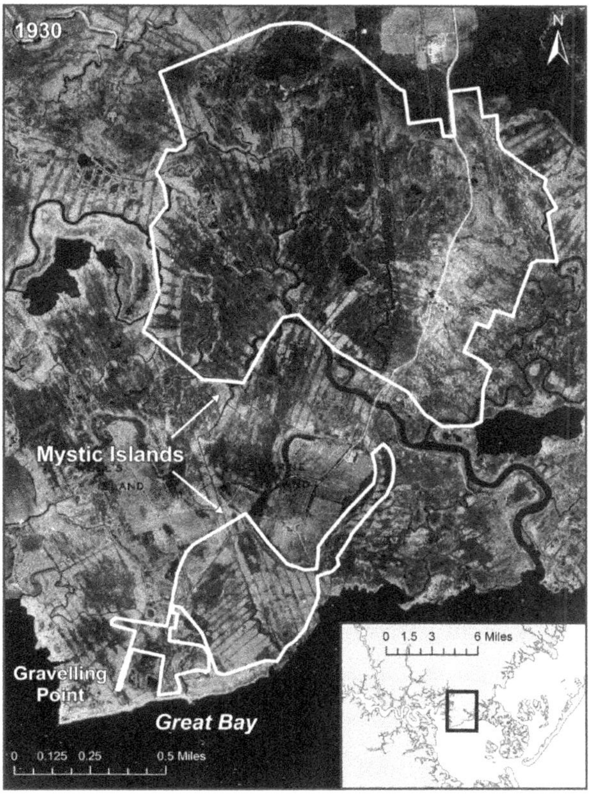

Figure 3.7 Mosquito ditches on the Sheepshead Meadows prior to development of Mystic Islands.

1928	• NJ proposes "Ocean Highway" including "Tuckerton Cutoff"
1929	• Road (S4-A) construction begins
1930	• Postponed construction of bridges
1931	• Electric power to the peninsula
1932–1937	• Road under construction with five temporary bridges
1938	• Further delay in construction
1940	• Road closed
1944	• Repair of road, construction of three bridges, repair of two bridges
1947	• State Highway S4-A renamed Great Bay Boulevard
1950	• Ocean County, Little Egg Harbor Township and Tuckerton, repairing bridges and roadway
1972	• Rutgers University purchases the old Coast Guard Station from the federal government
1976	• NJDEP completes purchase of Sheepshead Meadows and forms Great Bay Wildlife Management Area
1980s	• Seven Bridges Road closes for repairs
1986	• Arson at bridge over Big Thorofare
1989	• Bridges at Little and Big Sheepshead Creeks closed
1992	• Bridges over Little and Big Sheepsheads Creeks reopened
1994	• Bridge over Big Thorofare completely repaired
Mid 1990s	• Road repaired
Fall 2012	• Road by bridge over Big Thorofare repaired after damage in Hurricane Sandy

Figure 4.1 Timeline for planning and construction of Seven Bridges Road (Great Bay Boulevard).

Chapter 4
The Road to Nowhere

The varied history of Seven Bridges Road over nearly a century (Fig. 4.1), serves as a portal for viewing how this portion of the coast developed. This history included a planned, but never completed, major highway to and beyond this relatively remote location, a site representative of New Jersey's "low country." This chronology tracks the initiation of the road to the present, where, today, it has become a road to somewhere. Through time, the road carried many names, but "Seven Bridges Road" is that most common moniker residents and visitors to the area use as an identifier.

Likewise, during the same period, parts of the Sheepshead Meadows, the adjacent uplands, and the adjacent islands have carried a variety of names (Fig. 4.2). Some have particular reference to the local topography, such as points of land along the water (Graveling Point, Marshelder Point, Whirlpool Point) or capes (Cape Horn), thorofares (Big and Little Sheepshead Creek thorofares), specific creeks (e.g. Jimmies Creek [a thorofare], and coves (Horsefoot Cove). Other names are frequently used by local fishers in the area (No Name Creek, Go Through Creek, Shipwreck Creek). Of particular interest is the appellation "Sheepshead," as used in Sheepshead Meadows, Big Sheepshead Creek, and Little Sheepshead Creek. The sheepshead fish was apparently quite common and important in the area during the 1800s and early 1900s because it served as a white food fish with easily avoided large bones. With a range from Nova Scotia to Florida, its prior abundance likely caused the naming of Sheepshead Bay in Brooklyn, New York. Commercial fisheries around Woods Hole, Massachusetts, and in southern Chesapeake Bay caught large quantities of sheepshead in the early 1900s. It was so popular that fishermen at the beginning of the twentieth century indicated that they would rather catch one sheepshead than fifty monkfish. It declined in all these areas, however, after the first few decades of the early 1900s. Sheepshead fish all but disappeared around 1910, but the name remains in the landscape. It became virtually unknown in New Jersey waters, as demonstrated through our own extensive collecting with

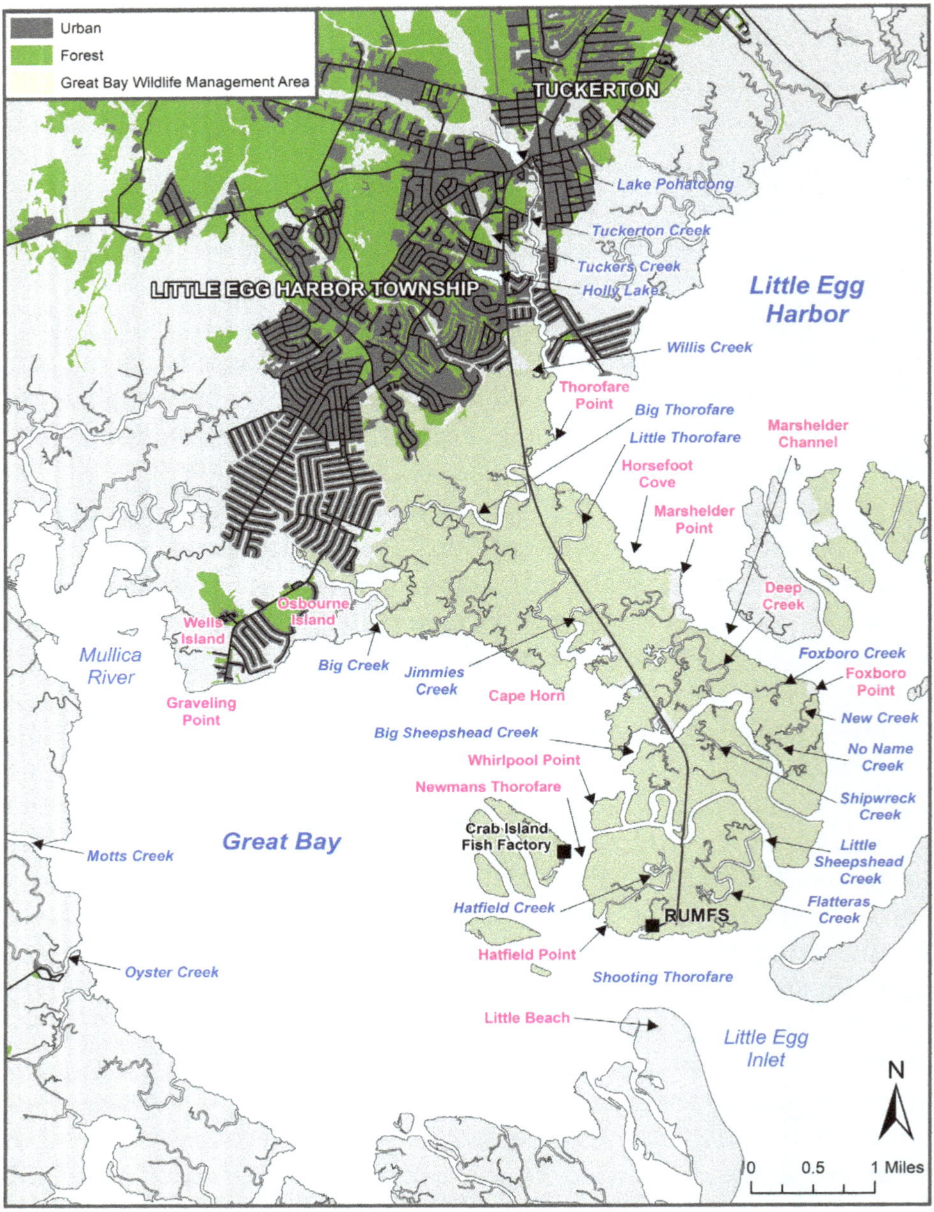

Figure 4.2 Official and unofficial names applied to the Sheepshead Meadows, adjacent uplands and islands.

a variety of sampling gears. Within the last two decades, sheepshead have returned to the waters around the Sheepshead Meadows. Since 2017, a number of individual fish have been collected, and in September 2022, a state record fish of 13 pounds, 3 ounces, was speared off the coast.

The Planning Process and Early Construction

In January 1927, the State Highway Department proposed a comprehensive renumbering of state highway routes. The legislature accepted the proposal and memorialized it into law two months later. The law included Route S-4-A, "extending from Route 4 at or near Tuckerton and extending to a point on Little Beach; provided, however, the county of Atlantic shall first agree to construct a suitable continuation of said road from Little Beach to the city of Atlantic City." In reviewing period newspapers, the first mention of the road occurs in 1928, reporting on the law passed the year before. The article indicates the State of New Jersey proposed a road connecting Tuckerton, and the rest of the mainland, to Atlantic City and other barrier islands in the chain via an "Ocean Highway." The local connection from Tuckerton became known as the "Tuckerton Cutoff." In general, the plans called for building a road extending from the mainland across the Sheepshead Meadows to Little Beach and Brigantine to Atlantic City. This route called for five bridges across the Sheepshead Meadows and a sixth bridge to carry the lower end of this road across Shooting Thorofare to Little Beach Island, down through Brigantine, plus another span to Atlantic City, hence the name "Seven Bridges Road" (Fig. 4.3). This approach was believed to shorten the trip from the mainland to Atlantic City by 10 miles. Another planned bridge, part of the proposed "Ocean Highway," would connect the end of the Tuckerton Cutoff to Long Beach Island. Some accounts suggest that the road and bridges could be completed in 1930 or 1931.

Construction actually began in 1929 with filling and profiling the roadbed. One contractor began at the north end near Tuckerton, and a second contractor worked from the south end near Shooting Thorofare (Fig. 4.4, 4.5). As one sign of the activity, the contractors rented a building near the Tuckerton end of the road for the "colored" employees who would be working on the road. An undated photo shows gas pumps used to fill construction vehicles near the end of the road (Fig. 4.6). Construction problems quickly arose. Midway through 1930, workmen began constructing three temporary bridges to facilitate work crews, but a lack of funds delayed the work, although the roadbed itself and some of the temporary

Figure 4.3 Early illustration of the southern portion of Long Beach Island with the proposed Seven Bridges Road between Tuckerton and the Atlantic shore. Note that five bridges cross from the mainland to the barrier islands but other bridges to Brigantine and Beach Haven Heights are also indicated.

bridges are visible through the Sheepshead Meadows about this time (Fig. 4.7). The bridge over Shooting Thorofare does not appear and it was never constructed, likely due to Atlantic County's lack of funds during the Depression to meet its legislatively mandated obligation to build the remaining two bridges. Thus, the moniker "Road to Nowhere." By late 1938, with the Great Depression grinding on, the likelihood of further funding for the road became even more remote.

In effect, the creation of Seven Bridges Road across the Sheepshead Meadows formed a low-level dam through the marshes, except at the five

Figures 4.4–4.6 (Upper left) Initial stages of construction of Seven Bridges Road in Tuckerton in 1929. Note the Old Bartlett House in the upper left which is now part of the Tuckerton Seaport. (Upper right) Fill being added to Seven Bridges Road during initial stages of construction in 1930. Photo for 4.4 and 4.5 by Marian V. Leake from the Tuckerton Historical Society. (Bottom) Photo of gas pumps along the side of Seven Bridges Road that were used during construction through the Sheepshead Meadows. Photo from the collection of John Bailey Lloyd.

bridges, where water could freely flow and vessel traffic could pass. To provide for additional flow through this "dam," workmen installed culverts under the road. Today, these culverts remain in situ, at seven locations between Tuckerton and Shooting Thorofare. A number of these culverts connect small creeks that once flowed freely where the road now exists (Fig. 4.8). The culverts are variable in number, size, and arrangement, with as many as three pipes running together under the road as at the unnamed creek near Captain Mike's Marina (Fig. 4.9). At another location, the presumed culvert arrangement that used to drain a marsh area has now been replaced with a small dam, constructed between 1956 and 1963. This dam allowed the formation of present-day Holly Lake (Fig. 4.10). Another location features a larger constructed culvert support structure where it runs under the road (Fig. 4.11).

Figure 4.7 Aerial image of construction of Seven Bridges Road in early 1930s between Route 9 in Tuckerton to end of road at Shooting Thorofare. From Historic Aerials.

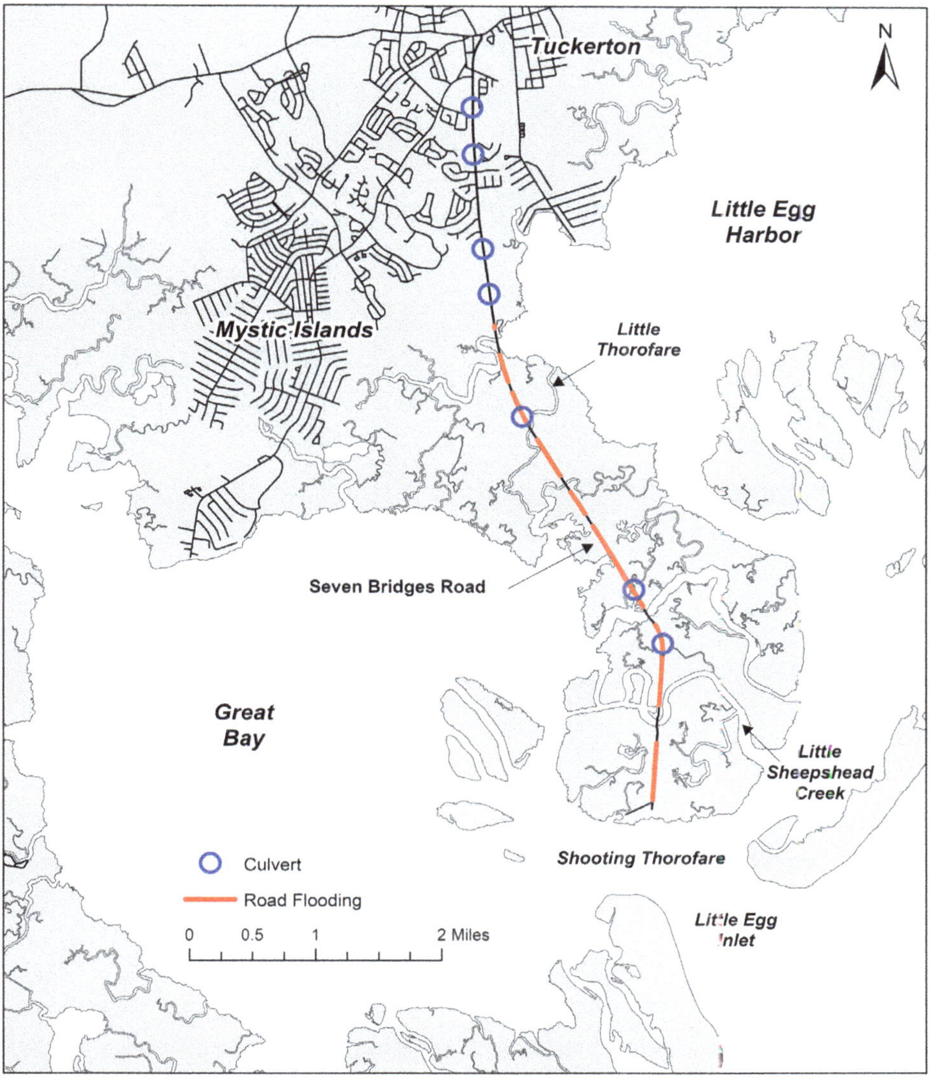

Figure 4.8 The location of culverts along Seven Bridges Road and an indication of the portions of the road that are frequently flooded. The culverts, in sequence from Tuckerton to Shooting Thorofare at the end of the road are those at 1) Tuckers Creek that drains the Pocket Marsh near Oak Road, 2) that drains Holly Lake into a lagoon development on the opposite side of the road, 3) Willis Creek that drains from the marshes and the lagoon development on the one side to the remnants of the creek on the other side, 4) adjacent to the Tuckerton Mound where there are marshes on both sides of the road, 5) a similar area embedded in marshes just west of Little Thorofare, 6) a larger unnamed creek just north of Captain Mikes Marina, and 7) one that drains from the marsh on the west side to a small unnamed intertidal creek on the opposite side. Map prepared by the author and Ryan Larum.

Figure 4.9 Three culverts providing water flow in a relatively large, unnamed creek near Captain Mikes Marina.

Subsequent Problems and Solutions

During the Prohibition years (1920–1933) "rumrunners" used this isolated road as a port of entry into New Jersey. They brought illegal booze by boat, often from larger vessels offshore, to the remote end of Seven Bridges Road, including Big and Little Sheepshead creeks, and offloaded it into trucks for delivery to the mainland and the nearby railroad. As evidence of the importance of this location, the Coast Guard from the Bond Station on Long Beach Island captured boats with illegal booze in Great Bay off Oyster Creek in October 1924 after raking the rumrunners boats with machine-gun fire. In another newspaper account from December 1931 a rumrunner ran aground in Little Egg Inlet with $25,000 worth of illegal liquor. Even though Prohibition was repealed in 1933 the high price of liquor fostered continued illegal activities by rumrunners until circa 1936. A lonely spot on Seven Bridges Road was where Jake Knove, the Atlantic City bootlegger boss with numerous aliases, was murdered with six bullets to the head in 1936. It was about the same time that the U.S. Coast Guard purchased from

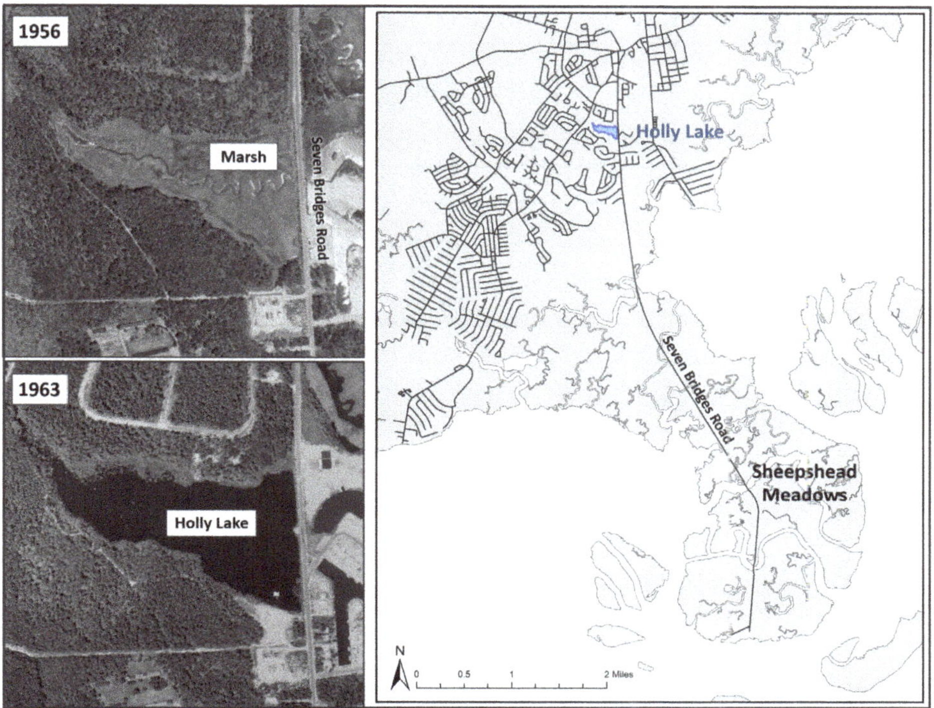

Figures 4.10–4.11 (Top) Holly Lake was formed between 1956 and 1963 with the construction of a dam that resulted in the inundation of the former marsh. Image from Historic Aerials. (Bottom) The larger culvert support structure as it presently exists where the remnants of Willis Creek flow under the road.

the Joseph Wharton estate, then the owner of the fish factories on Crab Island, the property adjacent to Willit Creek for a temporary Coast Guard facility and eventually a new station, constructed in 1937.

In 1939, Ocean County, Little Egg Harbor Township, and the town of Tuckerton (for the section closest to the town) supported and approved the reopening of the road, still referred to then as State Highway S4A. Justification for completing the roadway stemmed from the need for road access to the recently completed Coast Guard Station near Shooting Thorofare, the several businesses established along the route, and the fact that "thousands of sportsmen continuously use this road to go to and from the famous fishing grounds of Little Egg Harbor Bay" (an old name for Little Egg Inlet and Great Bay). In 1940, these visitors issued "a wave of public protest" because the road contained "a number of row boat, bait, restaurant and other types of business establishments" that were no longer accessible. At about the same time, government officials proposed relocating a section of the Intracoastal Waterway, taking it on a more westerly route through southern Barnegat Bay, including through the Sheepshead Meadows (Fig. 4.12), but nothing resulted from this plan. In 1941, the portion of today's Seven Bridges Road was quietly removed from the state highway system matrix.

By 1944, efforts succeeded in completing the road, based on plans for "refilling and resurfacing of a large portion of the road bed for nearly six miles, construction of three bridges and repairs to two small bridges." This included the driving of test pilings at Big and Little Sheepshead Creek bridges and plans for the bridge over Little Thoroughfare. Officials expected to complete the entire project by the end of the year. By 1947, the road became officially known as Great Bay Boulevard, and this formal name persists today. Contrary to the optimism expressed in 1944, some of the bridges were finally completed in 1948. Much of the end of the road, however, comprising the surrounding meadows, remained part of the Crab Island fish factory property. During this period, Atlantic City Electric extended power lines along the road. By 1950, officials planned to keep the road open during the summer season by repairing the bridges and roadway. Efforts to provide a larger bridge over Big Thorofare delayed this plan.

As the fish factory on Crab Island grew into a small city, so did adjacent developments on nearby Seven Bridges Road. These included a weigh station for offloading menhaden reduction products from the Island (Fig. 4.13). At the same time, the intense activity at the fish factory resulted in an increase in the number of connections from the "suburbs" at Little

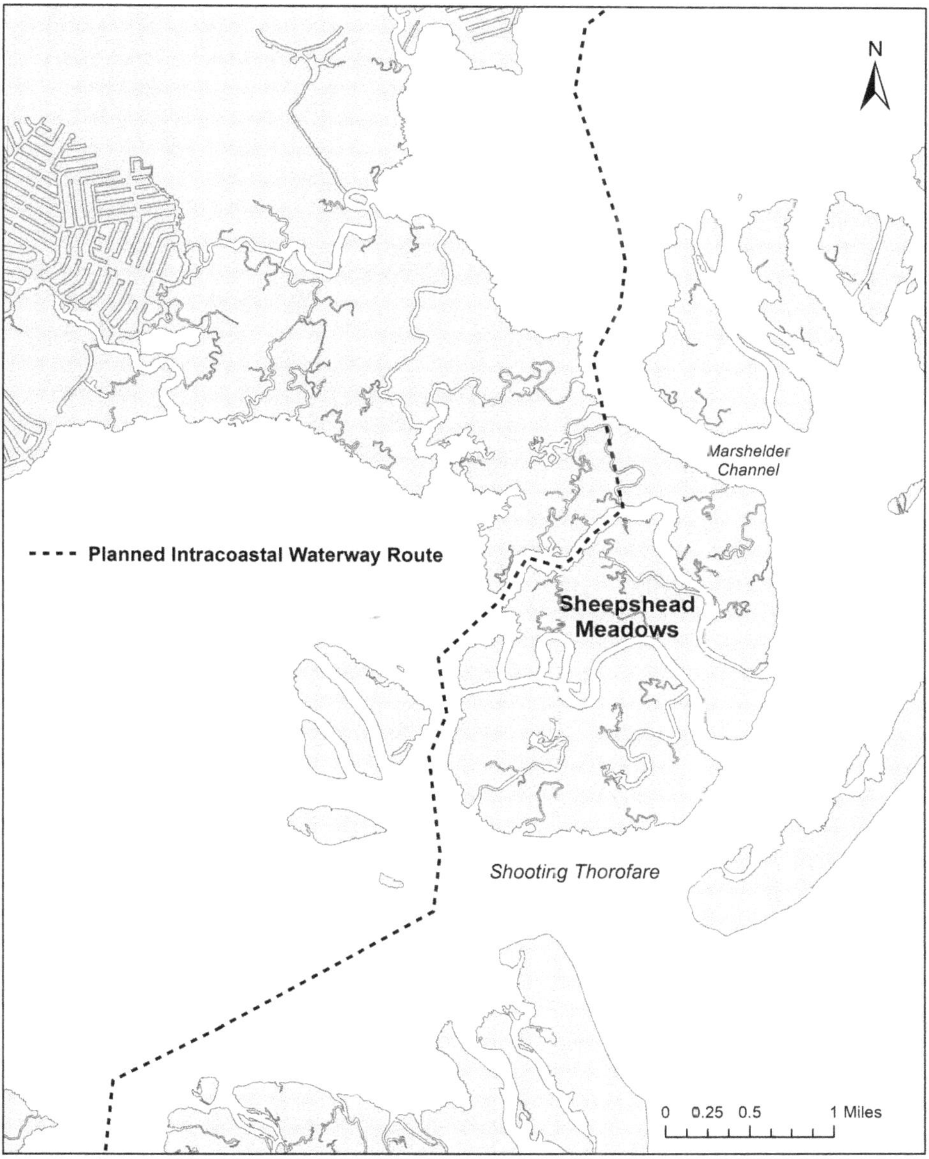

Figure 4.12 The planned route for the Intracoastal Waterway was western Little Egg Harbor through the Sheepshead Meadows and into Great Bay. This route would have required dredging through the meadows into Big Sheepshead Creek to make a connection through Newmans Thoroughfare. See Fig. 1.1 and 1.2 for locations.

Figure 4.13 Location on Seven Bridges Road for the measuring and transfer of products from the Fish Factory on Crab Island. The products were offloaded from boats/barges into trucks, weighed on the larger scales that used to exist in the building and then delivered to their various markets. Photo supplied by Tuckerton Seaport.

Sheepshead Creek, where docking facilities on Seven Bridges Road along the creek (Fig. 4.14) extended to the bridges over the creek (Fig. 4.15). By 1964, the Coast Guard had abandoned Station 119 as part of an overall effort to reduce the number of manned stations along the coast.

The Road to Somewhere

Several events during the 1970s enhanced the importance of Seven Bridges Road. In 1972, Rutgers University purchased the abandoned Coast Guard Station No. 119 at the end of the road from the federal government. The station's repair and renovation began to adaptively reuse the building as the Rutgers University Marine Field Station through the Center for Coastal and Environmental Studies, and later the Department of Marine and Coastal Sciences. Much of these early efforts focused on the repairs necessary from a fire that damaged the main building in 1969. Further stabilization of the road and bridges for Seven Bridges Road began with the purchase of 4,670 acres from J. Howard Smith, the owner of the fish

Figure 4.14 Image of the bridge over Little Sheepshead Creek where several docks with vessels provided an additional facility for transferring people to and from the Crab Island Fish Factory. Photo supplied by Tuckerton Seaport.

factory on Crab Island, and much of the Sheepshead Meadows. With the sale completed in 1976, the area became known as the Great Bay Wildlife Management Area, thanks to the efforts of Pete McLain (Fig. 4.16) in his roles as Federal Aid Coordinator Planner and Deputy Director of the New Jersey Department of Environmental Protection Division of Fish and Game. Ocean County, Little Egg Harbor Township and the town of Tuckerton retained the repair and maintenance of the access road, Seven Bridges Road. Since then to the present, the area continues to be a hot spot for waterfowl hunting, birdwatching, fishing, crabbing and clamming. Later, the Nature Conservancy assisted in acquiring an additional tract of salt marsh and coastal plain uplands as an addition for a total of 5,357 acres.

Despite these improvements to the area, the remote location has fostered other problems. Soon after the repair of the three bridges closest to Tuckerton, the nefarious work of an arsonist heavily damaged the closest bridge in 1986. This bridge underwent temporary repair using a "Bailey Bridge" that permitted limited access. Shortly thereafter, in 1989, the county

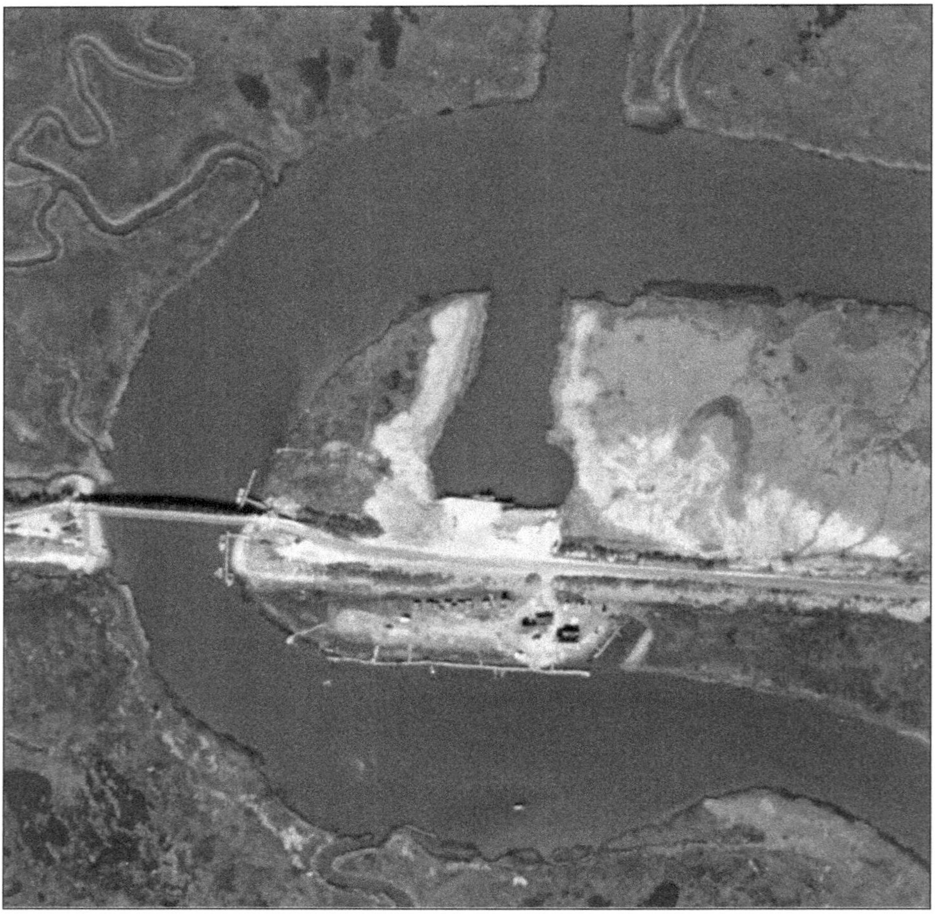

Figure 4.15 Aerial image from 1957 showing the many activities associated with the fish factory on nearby Crab Island including man-made boat basin and a variety of docks along the perimeter of Little Sheepshead Creek and Seven Bridges Road. From Historic Aerials.

engineering department closed the old bridges at Little and Big Sheepshead creeks to vehicle traffic for safety reasons (Fig. 4.17). The road suffered further deterioration (Fig. 4.18) and did not fully reopen until 1992. Then, in 1994, the bridge closest to Tuckerton reopened completely. During the mid-1990s, the road received new paving, which made vehicular travel much more accessible to the general public, relative to previous decades. These changes also improved access to the historic marinas along Seven Bridges Road including Captain Mikes (Fig. 4.19), Cape Horn Marina (Fig. 4.20) and the former Rands Marina (Fig. 4.21), along with other parts of

Figures 4.16–4.17 (Top) Pete McLane, closest to Great Bay Wildlife Management sign. Joining him are others who contributed to our enhanced understanding and use of the area: Bob Birdsall (center) and Terry O'Leary (right). (Bottom) Status of the two wooden, single lane bridges at the southern end of Seven Bridges Road is reflected in the bridges closed to vehicle traffic. Rutgers University Marine Field Station personnel had to ride bikes to get to the Station. Photo by Rutgers University Marine Field Station staff.

Figures 4.18–4.19 (Top) The poor status of Seven Bridges Road, with numerous potholes, during the early 1990s. The standing joke by Rutgers University Marine Field Station personnel during the time was that the potholes were not that deep because we dove in them. (Bottom) Image from helicopter of Captain Mikes Marina along Seven Bridges Road in the Sheepshead Meadows with bridge over Big Sheepshead Creek during May 2017 (see inset for general location). Note the Crab Island fish factory on Newmans Thorofare in the upper left and Great Bay in the distance.

Figures 4.20–4.21 (Top) Image of Cape Horn Marina along Seven Bridges Road. The covered structures at the base of the facility closest to the road are Bayfarm Bivalve Mariculture operation that takes advantage of the clean water from adjacent Great Bay. Drone image by Rob Auermuller. (Bottom) Image from helicopter of the former Rands Marina (now part of the Great Bay Wildlife Management Area) along Seven Bridges Road in the Sheepshead Meadows with Great Bay in the background (see inset for general location).

Figures 4.22–4.23 (Top) Public landing and fishing dock on Seven Bridges Road at First Bridge. Drone Image by Rob Auermuller. (Bottom) Image from a helicopter of the end of Seven Bridges Road during 2021 (see inset for general location). Note the extensive unaltered, meandering creeks through the abundant marshes of the area. The bridge over Little Sheepshead Creek is evident (upper center).

the remote area. At the same time, First Bridge Marina closed, which the State of New Jersey purchased and turned into a fishing dock for the general public at the bridge closest to Tuckerton (Fig. 4.22). Despite these human modifications to the area, much of it remains an undisturbed marsh with numerous meandering creeks (Fig. 4.23).

At about the same time, extensive additions to the Rutgers University Marine Field Station occurred in Tuckerton with the completion of an annex to the Station buildings near the end of the road (Fig. 4.24). These expanded facilities included a large dormitory, warehouse, and storage yard. As part of this expansion, the field station added new larger coastal vessels, like the R/V (Research Vessel) Arabella. An image of this vessel recently appeared on a local brewery's containers (Fig. 4.25). These new research activities also included establishing an extensive meteorological array staffed with Rutgers University personnel at the former Crab Island fish factory weigh

Figure 4.24 Helicopter image of Rutgers University Marine Field Station Annex (dormitory, upper building with red roof, and warehouse) and JCNERR facility (lower building with red roof) along Seven Bridges Road and adjacent to Fred and Judy Grassle Pocket Marsh with white boardwalk extending into the marsh surrounding Tuckers Creek (see inset for general location).

station site along Seven Bridges Road at Little Sheepshead Creek (Fig. 4.26). As a result of all of these activities, the Rutgers University Marine Field Station "graduates" have moved on and now occupy positions along the east, west, and gulf coasts of the United States, as well as internationally.

As a further indication of the Rutgers University Marine Field Station's importance, more outreach activities have become commonplace, including filming a kayaking video in the Sheepshead Meadows (Fig. 4.27). Holding an Open House in the fall of each year in the last several decades provides one way to inform the general public about our research and learning opportunities (Fig. 4.28). These annual public events serve as the ideal

Figures 4.25–4.26 (Top) An image of the R/V Arabella in the Rutgers University Marine Field Station boat basin was captured on the containers of a local brewery. Photo by Roland Hagan. (Bottom) Image of the Rutgers University Department of Marine and Coastal Sciences meteorological station just before the bridge over Little Sheepshead Creek along Seven Bridges Road. Drone Image by Rob Auermuller.

Figures 4.27–28 (Top) Filming of a video of a kayak trip in the vicinity of the Rutgers University Marine Field Station (RUMFS) at the end of Seven Bridges Road. (Bottom) A typical day at the fall Open House at RUMFS. This view is of some of the displays on deck while others occur along the causeway to the station and inside the main building.

Figure 4.29 One of the displays, devoted to visitors of all ages, led by graduate student Jessica Valenti while she explains how the ear bones (otoliths) of fishes can be used to provide information on their age, growth and habitat use relative to salinity.

setting for communicating our estuarine research (Fig. 4.29). Over the years, thousands of people of all ages from New Jersey and adjacent states have attended (Fig. 4.30). In 1997, the Jacques Cousteau National Estuarine Research Reserve was established with buildings immediately adjacent to the Rutgers University Marine Field Station Annex (Fig. 4.31) on Seven Bridges Road close to Tuckerton. The Cousteau Reserve has dramatically enhanced public outreach—to municipalities in general throughout the area and the rest of coastal New Jersey.

These numerous additions and improvements along Seven Bridges Road, especially those related to research, teaching, and outreach relative to estuaries and marshes, have fostered enhanced use of Seven Bridges Road since the 1970s. As a result, we, in collaboration with numerous fishers, birders, and naturalists, have developed numerous insights into these topics as covered in the following chapters. All is made possible through access to the natural habitats found on Sheepshead Meadows and the adjacent islands. Because of these abundant natural habitats, relatively unaffected

Figures 4.30–31 (Top) A display about New Jersey fish and crab fisheries by Doug Zemeckis of Rutgers University Cooperative Extension on the Rutgers University Marine Field Station porch overlooking the boat basin, during the annual Open House. (Bottom) The Jacques Cousteau Estuarine Research Reserve building is located on Seven Bridges Road in Tuckerton. Photo by Bob Birdsall and Terry O'Leary.

by human alterations, the record setting flooding by Superstorm Sandy had relatively little impact on the Sheepshead Meadows and the adjacent islands, unlike the nearby populated areas of Mystic Islands and Tuckerton Beach. Covid-19 restrictions further enhanced human use of the Sheepshead Meadows through naturalists, fishers, crabbers, hunters, cyclists, dog walkers, and adventurers using the road simply because people grew weary of being isolated inside during the period from 2020–2022 (Fig. 4.32).

Figure 4.32 The bridge over Little Sheepshead Creek provides access to strong tides and deep water and thus is often full of fishers during spring through fall. Be aware that these make car traffic slow and the light may change before completing the crossing.

Chapter 5
The Modern Coast

Several events influenced the modern New Jersey coast formation, although much of that bypassed Seven Bridges Road and the surrounding islands. The end of World War II brought many changes, including a robust economy and more recreational time. Construction of the Garden State Parkway was among the most influential events for the New Jersey shore. Its completion prompted a human migration to the shore, both seasonal and permanent. This in turn led to numerous changes along the shore, including lagoon developments such as Mystic Island and Osborne Island, along with the modernization at the menhaden fish factory on Crab Island. The creation of the Great Bay Wildlife Management Area, which protected this low country from further human modification, prevented developmental expansion into the Sheepshead Meadows. In subsequent years, this protected area provided the perfect setting for developing the Rutgers University Marine Field Station and the university's fiber optic link through Little Egg Inlet, investigations from other universities, as well as the formation of the Jacques Cousteau National Estuarine Research Reserve. Together these facilities have prompted numerous research, education, and outreach activities that stretched to the entire coast of New Jersey and beyond.

Progress and Failures in the Menhaden Industry

The industrial fishery for Atlantic menhaden operated with purse seining in the ocean (Fig. 5.1–5.2) and then processed the catch at land-based fish factories. The industry associated with the menhaden fishery constitutes one of the longest-lived, resource-based industries in the region (Fig. 2.9). It persisted as long as it did, despite two world wars, destructive fires (including the fish factory on Story Island in 1938), and the continued availability of menhaden, because of the adaptations and changing markets in the fishery. In 1935, J. Howard Smith purchased the former fish factory and garbage dump on Crab Island (and much of the surrounding Sheepshead Meadows). He and other members of his family began processing

Figures 5.1–5.2 (Top) Seine boats hauling in the purse seine full of menhaden in the ocean. Once the catch was concentrated they were pumped, in the more recent years, on board the mother ship for transportation to a fish factory on shore such as that on Crab Island. Photo by Joe Smith. (Bottom) Hauling back menhaden for processing. Photo from Tuckerton Seaport. Both images circa 1950s.

menhaden again. In the course of rebuilding, the modified factory (Fig. 5.3) incorporated advances in the production of fish meal and fertilizer. Smith made better use of the bycatch of larger fish (weakfish, bluefish), all caught from larger vessels using purse seine nets, and selling them to fishmongers. Other enhancements included a water supply, with a water tower, that Smith constructed on Crab Island in 1947 (Fig. 5.3). At about the same time, Smith began employing radio-equipped spotter planes to locate menhaden schools in the ocean. The pilots would then direct the fishing boats to catch the large schools. The remnants of a linear landing strip built on Crab Island reflect this development.

Between 1944 and 1950, Smith had a new channel dredged from the fish factory through the meanders in Little Sheepshead Creek (Fig. 5.4) to a new facility on Seven Bridges Road, where workmen loaded products from the fish factory on trucks and then weighed for transportation to various markets (Chapter 3). At the peak of factory activity, estimates indicate that up to 200 workers lived in the "city" on Crab Island or commuted from Tuckerton and

Figure 5.3 The painting "Black Ducks at the Fish Factory" was the cover for the Third Annual Old Time Barnegat Bay and Gunning Show Program in 1992. It depicts the fish factory when it was at its heyday in the early 1950s.

nearby areas via tugboat between the road and the island from the dredged basin or a small dock off of the bridge over Little Sheepshead Creek. The nature of the products changed, especially late in the life of the fish factory in the 1960s, when the factory produced high protein fish flour for a variety of markets from chicken food to fish food for aquaculture. Eventually the Crab Island fish factory closed in 1965, when the stocks of menhaden in the ocean declined. In 1982, and again in 1991, major fires burned much of the fish factory, its piers, as well as many of the cottages which once housed the factory workers (Fig. 5.5, 5.6). The New Jersey Department of Environmental Protection subsequently purchased Crab Island and the adjacent property on the Sheepshead Meadows and other islands and the area became the Great Bay Wildlife Management Area (Chapter 3). A fire in November 1993 destroyed the former fish factory weigh station on Seven Bridges Road. Rutgers University now leases the site and installed an array of meteorological instruments for monitoring the estuary and coastal ocean. This has proven exceptionally valuable for oceanographic and estuarine research, in general, and evaluation of wind energy in the nearby coastal zone.

Figure 5.4 Aerial image of the dredged channel from the fish factory on Crab Island (upper left) through the meanders in Little Sheepshead Creek to Seven Bridges Road and the dredged boat basin at the site of the weigh station on the road from May 2017.

Figures 5.5–5.6 (Top) Aerial image indicating the remnants of the numerous building for the menhaden processing as well as housing for the workers (behind and to the right of the water tower) on Crab Island. The linear feature visible from just above the top of the water tower is the former landing strip for the spotter planes. From *Weird New Jersey* as published in the *Tuckerton Beacon*. (Bottom) Aerial image from 2015 indicating the loss of piers, with just the pilings remaining around the periphery. Most of the buildings are gone, due to fires, but the water tower and storage tanks remain.

Gunning Traditions

The marshes in the Sheepshead Meadows and the adjacent islands have served as hunting grounds since the time of the Lenape. It continues today, based on the seasonal migrations of waterfowl and shorebirds along the Atlantic Flyway. The frequency of hunting increased after European settlement and the explosion in human populations along the New Jersey shore and this is no doubt true for the Sheepshead Meadows as well. In the early days of gunning for food, as exemplified in an article on shorebird shooting on the flats near the mouths of the Sheepshead creeks published in in the late 1800s, almost all birds were fair game, including curlew, plover, yellowlegs, dowitcher, willet, and clapper rails. These birds were incredibly abundant in the 1800s and early 1900s, but continued gunning pressure, from dawn to dusk over most months of the year, caused a marked decline. Compounding this decline, an eelgrass wasting disease and drought struck in the mid-1930s. During this same period, many coastal marshes suffered loss for housing developments and other activities, although this was less evident in the Sheepshead Meadows and the adjacent islands.

Over time, there was a transition from individuals gunning for food in blinds, sneakboxes, sinkboxes and even floating blinds. Gunning became more common with groups of people, and houseboats frequented the marshes. At one extreme, small shanties became residences like the one that Ray Kennedy occupied along Seven Bridges Road just south of the bridge over Big Sheepshead Creek. Today, the remnants are hard to identify because only a few pilings remain from that structure.

A time period for which we have reasonable records is when the fish factory on Crab Island was in full operation during the mid-1900s. At that time the fish factory owners also held title to thousands of acres on the nearby Seven Islands, as well as on the Sheepshead Meadows. They provided access to these marshy gunning grounds for hire and for friends, mostly pursuing waterfowl. At one time, most of the southern end of the Meadows was closed off and only those with permission could hunt there.

Notes from the Road—The Origin of Square Marsh Pools

Square marsh pools are not designed by Mother Nature. Of the hundreds to thousands of marsh pools present in the Sheepshead Meadows and the adjacent islands, only a very few have linear sides, based on aerial images with helicopters and drones and measurements from the ground. Square marsh pools are so strikingly different from natural pools that they quickly stand out. We have identified two of these in our aerial exploration over the last five years. These share several characteristics including general shape, substrate, and proximity to the road. The most obvious of these pools is at the eastern end of Seven Bridges Road (now Great Bay Boulevard), just after the causeway to the Rutgers University Marine Field Station and just a few feet from the road (Fig. 5.7). The larger pool's linear sides are quite obvious on three sides. Connected to this pool is another, smaller pool. A similar pool is located along the same road between Little and Big Sheepshead creeks, about 30 feet from the road, but separated from it by an unnamed intertidal creek (Fig. 5.8). The largest pool is linear on all four sides. Immediately adjacent is an L-shaped pool. Both clearly differ from the irregular shape in nearby pools from both images. Both of the large pools have soft, muddy bottoms and lack structure. A metal probe inserted into the bottom at multiple locations in each of these ponds did not meet any resistance within depths of 2–3 feet.

The source of these pools is clearly human in origin, but otherwise uncertain. We have considered that they could have been hunting blinds or boxes because a large portion of the southeastern end of the Sheepshead Meadows served hunters associated with the fish factory property on Crab Island and included much of the Sheepshead Meadows. Another possibility is that they constitute "borrow pits," dug during road construction. The most

Figure 5.7 Aerial image of a man-made pool with linear sides between the causeway to the Rutgers University Marine Field Station and the end of Seven Bridges Road (on left). Note how the linear shape differs from the irregular shape of the other pools. The longest dimensions measure 44 by 31 feet for the larger pool, and 10 by 20 feet for the smaller pool. Drone image by Thomas Johnson.

Figure 5.8. Aerial image of two man-made pools with linear sides located across an unnamed creek adjacent to Seven Bridges Road between Little and Big Sheepshead creeks. The longest dimensions measure 32 by 27 feet for the larger pool, and 23 by 13 feet for the smaller L -shaped pool. Drone image by Thomas Johnson.

likely alternative is that someone dug these larger pools to attract waterfowl and the smaller, nearly square or rectangular pools, provided sites for sunken blinds to hide hunters. On some occasions, these reportedly contained aluminum structures permanently sunk into the marsh surface and then pumped out prior to use. In addition, these unusually shaped pools may have been more common, but the marsh grass growing and overhanging the marsh pools, as commonly occurs, likely obscure their margins.

Later, established gun clubs built large permanent structures for groups of gunners and their guides, especially in more remote locations, becoming commonplace, especially around the islands south of Tuckerton (Fig. 5.9). Several of these gun clubs remain in operation today. Two clubs occupy Hither Island, including both Hester Sedge (Fig. 5.10) and Hither Island (Fig. 5.11) and another club operates on Drag Island (Fig. 5.12). Gatherings such as the annual Decoy Show in Tuckerton showcase the history

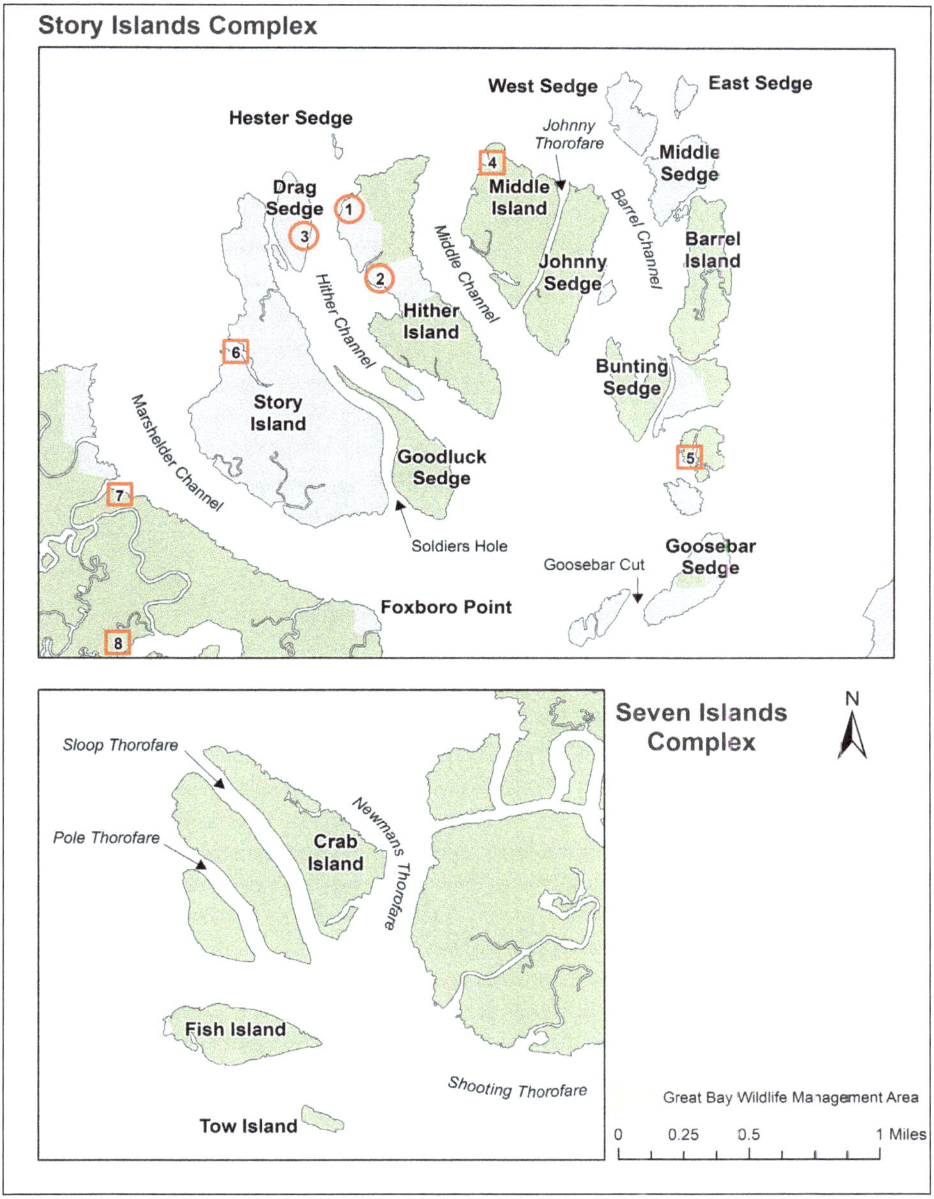

Figure 5.9 Locations of former and existing gun club buildings on the Sheepshead Meadows and the adjacent islands in Little Egg Harbor. Map by author and Ryan Larum.

Figure 5.10 (Top left) Aerial image of the present day (December 2021) Hester Island Gun Club at the edge of Hither Channel on Hither Island. The marsh in front of the building shows extensive erosion behind the old bulkhead that used to be at the shoreline. The marsh behind the building is dominated by extensive marsh pools and "lakes." Figure 5.11 (Top right) Image of the present day (December 2021) Hither Island Gun Club and adjacent bridge across an unnamed creek of Hither Island. The marsh behind the gun club building has abundant marsh pools and "lakes." Figure 5.12 (Bottom) Image of the present day (December 2021) Drag Island Gun Club on Drag Island at the edge of Hither Channel. The marsh in front of the building shows extensive erosion behind the old bulkhead that used to be at the shoreline. Drone images by Rob Auermuller.

of these clubs in the Sheepshead Meadows area, and the adjacent islands. The original "Hunting Shanty" (Fig. 5.13) can still be viewed, as can other displays at the Tuckerton Seaport. Among the displays at the latter location is the replica of the Hester Sedge Gunning Club building at the Seaport (Fig. 5.14). This building is characteristic of many clubhouses, featuring sparse, simple facilities that provided shelter from the elements between hunting forays, as well as a place to socialize in these natural environments.

Figure 5.13 (Left) The Hunting Shanty, the original building that preceded the Tuckerton Seaport, is now in the center of this facility. Drone image by Rob Auermuller. Figure 5.14 (Right) A replica of the Hester Sedge Gun Club in the Tuckerton Seaport from December 2021. Image by Rob Auermuller. Inside the building are photographic records of some of the gun clubs that existed in the area.

Shellfisheries

Some of the shellfisheries in the vicinity of Seven Bridges Road can be traced back to the time of the Lenape (Chapter 2), but they continue today, although with very different fishing techniques, modified from sustenance fisheries into recreational and commercial operations. They also vary from those for bivalves (oysters, hard clams) and snails (whelks), and for crabs. The earliest fisheries European settlers operated for bivalves harvested hard clams and oysters and accounted for Tuckerton being called Clamtown at one time. In these early days, the fishery employed techniques like treading, raking, or bull-raking. Beginning in the late 1800s, Rutgers scientists conducted extensive research on these important resources in the Sheepshead

Figure 5.15 (Left) Wheelbarrow full of seed clams at Bay Farms culture operation at Horn Point Marina on Seven Bridges Road. Figure 5.16 (Right) Clam culture lease site, as marked by stakes in Jimmies Creek, one of the thorofares passing through the Sheepshead Meadows.

Meadows and Little Egg Harbor area, and elsewhere in New Jersey. Since the early 1970s, however, the harvest at many of the fisheries is based on cultured hard clams from hatcheries such as the one at Horn Point Marina on Seven Bridges Road (Fig. 5.15). Once these seed clams are large enough the fishery deploys them in natural settings on staked lease grounds where they are often covered with netting to prevent predation by crabs and fish (Fig. 5.16). After growing for 1–3 years, they are harvested for a variety of markets. Despite the increasing reliance on cultured clams, there are occasionally good natural sets.

The fisheries for oysters also began with the Lenape (Chapter 2), but the location and techniques have changed dramatically, since the early 1970s, to an emphasis on harvesting cultured oysters. Some of this activity is occurring around the Sheepshead Meadows and the adjacent islands. These recent efforts include an emphasis on breeding disease resistant oysters by Rutgers scientists at the Haskin Shellfish Research Lab on Delaware Bay. Once the "seed" oysters are produced in hatcheries, they are caged in a variety of ways such as bagging on the bottom or in rafts. Some of this planting is happening out of the Love Oysters fishery out of Captain Mikes Marina on Seven Bridges Road (Fig. 5.17). Once they are deployed at sites such as the shallow areas of Great Bay, they require intensive manipulation

Figures 5.17 oyster culture operations (Love Oysters) out of Captain Mikes Marina with grow out sites around the perimeter of the Sheepshead Meadows.

to encourage growth for a variety of markets from the restaurant trade to wholesale markets. Since the location of these field culture operations are critical to fast growth and reduced predation, there has been an intensive effort to develop spatial planning tools such as that for Great Bay and Little Egg Harbor. One potential response of these culture operations in the vicinity of the Sheepshead Meadows is the occurrence, in recent years, of a natural set of oysters, including in the Rutgers University Marine Field Station boat basin on pilings, and a boat railway, as well as in adjacent Willit Creek, where sedge oysters are now common on the shells of ribbed mussels.

The fisheries for other molluscs are much more variable. In some years, there were extensive sets of blue mussels in the area closer to Little Egg Inlet where these wild populations are frequently harvested in deeper waters. These sets of small blue mussels can be especially evident at the end of Seven Bridges Road along Shooting Thorofare at the edge of the marsh in the spring. Most of these die in the warmer temperatures of the summer, however, and have been less evident with the warming water temperatures of recent years. A fishery for whelks has also developed, including in the waters around the Sheepshead Meadows as is indicated by the traps which sometimes wash ashore. The other major shellfish type that supports valuable commercial and recreational fisheries is a crustacean, the blue crab. This tasty species is harvested with baited pots (Fig. 5.18) in the spring through fall in creeks, thorofares, and the open waters of Great Bay or Little Egg Harbor or dredges closer to Little Egg Inlet during the winter.

Figures 5.18 The front of a blue crab, with eyes and antennae visible inside a crab pot as photographed in Willit Creek. Image by Matt Taynor and Zoe Zientek.

Power to the Peninsula

One of the most prominent man-made features on the Sheepshead Meadows is the power lines stretching along Seven Bridges Road and over the five thorofares with bridges. Atlantic City Electric first installed the poles and ran the lines in the early 1930s, at about the same time as the road's construction, to provide power to southern Long Beach Island from the mainland. To accomplish the connection, the utility installed cables under Shooting Thorofare, and potentially at the same time, under Newmans Thorofare to the fish factory on Crab Island (Fig. 5.19). An indication of the time since the construction is a "bridge" that crosses an unnamed tidal creek (Fig. 5.20). The base of the horizontal poles used to construct the bridge are buried on each side of the creek under more than one foot of sediment. The salt waters of the creek frequently inundate the center of the bridge, based on the ribbed mussels that are growing on the bridge deck. Later, in the early 1940s, after Coast Guard Station 119 was built at the end of the peninsula and the causeway to it completed, a power line was constructed to provide power there from the line along Seven Bridges Road. What isn't clear is what purpose the power line poles in the water parallel to the causeway served? Perhaps they supplied power to the temporary Coast Guard facility constructed at the end of Willit Creek, just outside the new Station.

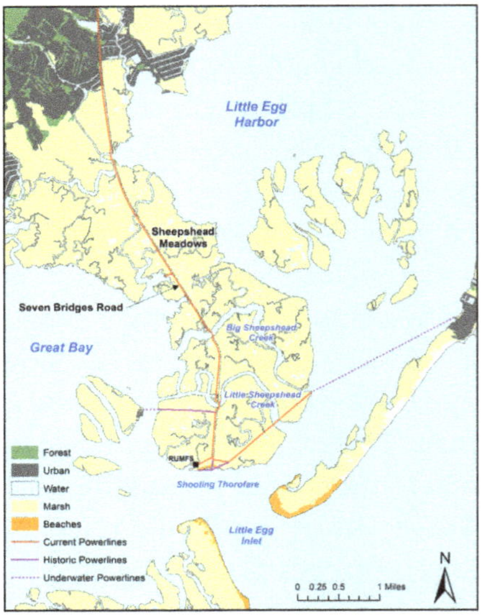

Figures 5.19 Arrangement of historic and current power lines along Seven Bridges Road including those underwater to Long Beach Island and the former fish factory on Crab Island.

Expansion of Recreational Fisheries

While sport or recreational fishing existed in the Sheepshead Meadows area prior to World War II, these expanded dramatically after that time. A major reason was more available leisure time, but in addition, improvements in boats (from wooden to lightweight aluminum and fiberglass and from wooden row boats to motorboats) and fishing equipment (lighter/synthetic rod and reels) prompted this change. Evidence for the increase in sport

Figures 5.20 "Bridge" over an unnamed creek presumably installed at the time of construction of the power lines from Seven Bridges Road to Long Beach Island in the 1930s from a photograph from December 2017. Note the eagles nest on one of poles. The power to the line was discontinued around 2004.

fishing can be found in the growing number of bait shops and marinas catering to the individual fishers, based on boat rentals and launch ramps, and the frequency of newspaper articles on the general topic. This was especially evident on Seven Bridges Road, where several marinas developed to serve the increasing number of those interested in sport fishing. At the same time, getting on the water became easier with the development of charter boats, such as the ones that fished out of Tuckerton Creek, and nearby Oyster Creek, and headboats out of nearby Beach Haven. Fish species of particular interest at the time included weakfish, Atlantic croaker, striped bass and bluefish (see Appendix 2).

Lagoon Developments

With the rise of leisure time and tourism after World War II, the development of shore-side housing became a prominent feature on the New Jersey coast, including around the Sheepshead Meadows. This is evident

Figures 5.21 Image from December 2021 of a large portion of the Tuckerton Seaport, with the Tuckers Island Lighthouse on Rt. 9, with Lake Pohatcong in the background, and many of the Seaport buildings on Tuckerton Creek in the foreground. Drone Image by Rob Auermuller.

beginning in the 1950s with the construction of lagoon developments off Seven Bridges Road and extending onto the Mystic Islands and Osbourn Island (Fig. 1.1 and Chapter 7).

Tuckerton Seaport

One of the most interesting, recent developments near the Sheepshead Meadows is the creation of the Tuckerton Seaport and Baymen's Museum, an area between Seven Bridges Road and Tuckerton Creek in Tuckerton (Fig. 5.21). This museum features extensive recreations depicting the rich history in the area, based on important activities of the last several centuries, including hunting, decoy carving, fishing, boat and ship building, as well as educational activities such as maritime forest walks and guided ferry tours along Tuckerton Creek and to nearby Beach Haven on Long Beach Island. The Seaport began at a small location, the Hunting Shanty, where the rich history of decoy carving was the central focus. Beginning in 2001 the Seaport eventually moved to its current location where it now includes replicas of numerous historic buildings, ranging from the former Tucker's Island Lighthouse, across the street from Lake Pohatcong, and the main building (Fig. 5.21). The significance of this location is evident at the

Figures 5.22 Another view of the Tuckerton Seaport from December 2021 with Rt. 9 and the Tuckers Island Lighthouse in the foreground. The dam spillway at Lake Pohatcon at Route 9 is in the lower left portion of the road. From this po nt Tuckerton Creek flows downstream into Tuckerton Cove and Little Egg Harbor (above and to the left). Seven Bridges Road is mostly hidden by the trees until it passes through the marsh and over the bridge at Big Thorofare (upper center). Drone image by Rob Auermuller.

top of Tuckerton Creek, as it stretches downstream to Little Egg Harbor Bay, and Tuckerton Cove to the Sheepshead Meadows (Fig. 5.22). One of the Seaport's advantages is the opportunity to view some of the recreated hunting shanties at gun clubs and houseboats in the region, especially at the islands near Story Island in Little Egg Harbor.

Figure 6.1 View of the southeastern portion of the Sheepshead Meadows from the cupola of the Rutgers University Marine Field Station. Extensive unaltered marshes, typical of the area, dominate the region with marsh pools and creeks.

Chapter 6
How We Gain Insights

Our insights into the Sheepshead Meadows have developed over decades, and at varied scales from the entire landscape to much smaller dimensions that can only be observed with a microscope or analyses of individual chemical elements. The "our" and "we" within this section comprises Ph.D. level scientists or faculty, and those in training at some level. The latter includes postdoctoral fellows, graduate students, interns, and technicians who participate in research. While the majority of these individuals are connected with Rutgers University Marine Field Station, and faculty from other Rutgers University campuses in New Brunswick, Camden, and Newark, there are many involved from other institutions. All of these scholars gain insights, prompted by a beneficial and thorough literature review focusing on the study area or comparable estuarine landscapes and habitats. This review prompts us to construct better questions and develop better answers that supplant or improve our previous knowledge base. This scientific process evolves slowly as we formally discuss ideas and issues, through a variety of avenues: at regional, national, and international meetings; at student's defense of their research; with land and fishery managers, and environmental groups. This winnowing process is critical to the scientific method as is our own activities and the products of our research.

The uniqueness and quality of the area under study, the Sheepshead Meadows and the adjacent islands, expedites gathering insights. Access to the six-mile-long Seven Bridges Road, through relatively unaltered salt marsh, and five (not seven) bridges, ramps at private marinas, and public landings, proves the value of "place" (Fig. 1.1, 1.7). Access to these sites is available both day and night and at various tidal stages. The Rutgers University Marine Field Station (Fig. 1.4) serves as the access nexus. This former Coast Guard Station, elevated above the marsh and waterways, accentuates all these characteristics, along with the visual (Fig. 6.1) and physical access it provides, even during storms. The islands adjacent to the Sheepshead Meadows fail to offer such access (Fig. 1.2).

Landscape Scale

On a broad scale, aerial imaging is obtained using helicopters and drones, and then these images undergo ground-truthing using Google Earth, boats, kayaks and wearing our boots (Fig. 6.2). These observations formerly extended beyond the Meadows to the adjacent ocean through a fiber optic link and a variety of sensors. This equipment operated for twelve-plus years and reached almost three miles into the ocean from the Rutgers University Marine Field Station to the LEO–15 (Long-term Ecosystem Observatory at 15 meters) data collection site, featuring an eighteen-square-mile well-sampled research space (Appendix 3). The LEO observation network also included multiple remote sensing, shipboard, autonomous, and moored sensor systems that surrounded a pair of instrument platforms secured to the ocean floor. As a result, it served as the impetus for other observation networks installed around the world.

The Jacques Cousteau National Estuarine Research Reserve in the estuary maintains other dataloggers. The one they maintain near the Sheepshead Meadows in the Intracoastal Waterway measures temperature, salinity, dissolved oxygen, pH, and turbidity at 15-minute intervals 24/7 since 1996. This equipment has also identified waterflow patterns at specific locations, such as near Little Egg Inlet, and for the whole estuary, including the waters around the Sheepshead Meadows and the adjacent islands. In addition, a variety of dataloggers have tracked and analyzed the air and weather over the Meadows and the adjacent ocean since 2002. As an example, the Rutgers

Figure 6.2 a–b Drones (left) and helicopters (right) have become a valuable part of our habitat research in recent years. Photos by author and Rutgers University Marine Field Station staff.

University Center for Ocean Observing Leadership (RUCOOL) operates a Coastal Metocean Monitoring Station on Seven Bridges Road near the Rutgers University Marine Field Station (Appendix 3, Fig. 6.3). This site houses a 40-foot meteorological tower which measures a variety of standard meteorological variables (such as temperature, dewpoint, and wind speeds). Additionally, a Triton SODAR (Sound Detection and Ranging) at the site uses sound pulses emitted to measure wind speed profiles up to 660 feet in height. A Lockheed Martin WindTracer scanning wind LIDAR system, which measures winds throughout the atmospheric boundary layer out to a range of six-plus miles, compliments the scientific equipment array. We can also virtually gather data on the aerial fauna such as birds (Fig. 6.3). Nitrogen deposition sampling provides analytical data as a eutrophication source adjacent to the former meteorological tower at the Rutgers University Marine Field Station. This approach, using an automatic precipitation sampler, was deployed and helped confirm the importance of atmospheric nitrogen inputs provided from urbanization, industry, and New Jersey agricultural sources and potentially even as far as the Ohio Valley.

Habitat Scale

We can easily observe changes in habitats across the landscape while traveling along Seven Bridges Road. The scale at which we make these observations occurs at different speeds, depending on modality, and thus influences the parts of the landscape and habitats we observe with how much detail. A car provides the most rapid passage. I estimate that during the 30-plus years spent commuting to the Station, I have driven this road over 18,000 times, always with my eyes and mind open to change that occurs over years, seasons, stage of tide, day versus night, and storms versus quiet times. At this speed, I observe extensive unaltered marshes and fringing red cedar forests and shrubs lining the roadway. Over the same time frame, what I can observe changes dramatically at a slower pace, as on a bike, providing the opportunity to distinguish different types of marsh from low, frequently flooded marsh to high marsh that is infrequently flooded and typically shorter. The powers of observation are significantly improved while walking, allowing us to see individual plants comprising the marsh and among the different shrubs that grow along the road. Remaining stationary, similar to birders, permits a more detailed surveillance, bringing stationary animals, details of flowering plants, as well as smaller insects and other invertebrates into view. In recent years, drones and helicopters

Figure 6.3 Varied approaches to understanding the Sheepshead Meadows and adjacent areas in the estuary and coastal ocean. Experimental marsh plots (center of upper left) with different degrees of flooding, draining or controls to evaluate these different regimes on the vegetation response. Telemetry antennae (upper right) as part of a coastal array for tracking tagged birds, with emphasis on federally threatened red knot and piping plovers and federally endangered roseate terns. This location is part of a collaborative effort led by University of Massachusetts, Bureau of Land Management, and U.S. Fish and Wildlife Service. Array of subtidal cages with small mesh for measuring the growth of juvenile fish over 10 day trials (middle left). Experimental marsh pool enclosure (middle right) being constructed to evaluate mummichog response in these habitats. Meteorological station (lower left) array along Seven Bridges Road. Blue crab on tether (lower right) to evaluate habitat and size-specific rates of predation.

served as stable platforms for aerial photography, dramatically increasing our understanding of habitats and their connectivity (Fig. 6.2).

At the air-water interface like the large intertidal areas that dominate the Sheepshead Meadows, we gain insights through a variety of sampling techniques for plants and animals (Fig. 6.3). These approaches include a variety of net types: dredges, but also trawls, gillnets, plankton nets, and traps (Fig. 6.4, 6.5). In deeper waters, we typically know less about the

Figure 6.4 A variety of nets used to sample larval and juvenile fishes, crabs and shrimps in the waterways in and around the Sheepshead Meadows. Larvae are sampled with small mesh plankton nets (upper left) while juvenile fishes can be sampled with pop-nets (upper right), seines (middle left), towed beam trawls (middle right), mini-seines (lower left), and small mesh gill nets (lower right). Photo by author and from the Rutgers University Marine Field Station archives.

Figure 6.5 Trapping of fish and crabs are frequent activities as with small mesh minnow traps (upper left), throw traps (upper right), experimental traps (middle left), light traps (middle right), pit traps (lower left), and oyster tray traps (lower right).

habitats, but Remotely Operated Vehicles (ROVs) with cameras or in situ video cameras have granted us new insights. The acoustic technology known as side-scan sonar allows us to detect structural features such as peat reefs: large marsh peat chunks that break off the edges and fall into channels. In deeper waters, structures exist that prevent the use of many net gears. In these instances, Dual-frequency Identification Sonar (DIDSON) offers important advantages in habitat structure and associated fauna identification verified by preliminary studies in the Rutgers University Marine Field Station boat basin and nearby Crab Island. Rutgers University scientists are pushing back the frontiers of sampling programs through developing the Navigator, a single drone that can transition from the air to underwater in two seconds. Testing this experimental drone has occurred at the Rutgers University Marine Field Station boat basin and beyond. In other instances, we use different tags for tag-recapture observations and underwater detection receivers to track the presence and movement of fish and crabs with acoustic tags (Fig. 6.6). A Stockton University group employs the same approach to track seasonally abundant harbor seals in Great Bay with satellite tags. Related observations include acoustic techniques for detecting surprisingly noisy marine mammals and reproducing fish (Fig. 6.7).

Notes from the Road—Faunal Sampling Techniques Make All the Difference

Marsh pools typically are small, enclosed, shallow habitats that should be easy to sample and determine the kinds, number, and size of fish that occur there. Despite these characteristics, these pools often prove difficult based on the available sampling gear. When we sampled with five different gear types during the day and the night, what we found depended on the equipment used. Some gears, including minnow traps, throw traps, and light traps had similar species composition (mummichop and sheepshead minnows) while other traps had different species such as for mini-seine (inland silversides, rainwater killifish) and light traps (all of the preceding including others such as Atlantic silverside, spotfin killifish). Fish size differed based on the gear used. Wire mesh traps and experimental traps caught fish over the widest average sizes (both up to about two inches) respectively. Other gears caught smaller fish including throw traps, light traps, and mini seines (about one inch). These gear type biases occur more commonly in open water habitats such as bays and creeks. Together, these kinds of observations make it clear that we need to maintain awareness of potential biases associated with different sampling techniques.

Figure 6.6 Images of a variety of tag types used for juvenile and adult fishes. Coded wire tag at end of a probe on a finger for tagging small fish (upper left). In this case for bluefish, they were held in small wading pools after collection and before tagging and release (upper right). Most of the individuals in the photo are interested onlookers that are provided impromptu lectures on the research. Ultrasonic transmitter used for tracking large fish after surgical insertion in the body cavity or attached externally (middle left). Striped bass that had just had an acoustic tag inserted in the body cavity at the point where the sutures and a yellow spaghetti tag was attached (middle right). These and other acoustically tagged fish were tracked to determine their location and habitat use (lower left). Small spaghetti tags inserted in a summer flounder muscle mass (lower right).

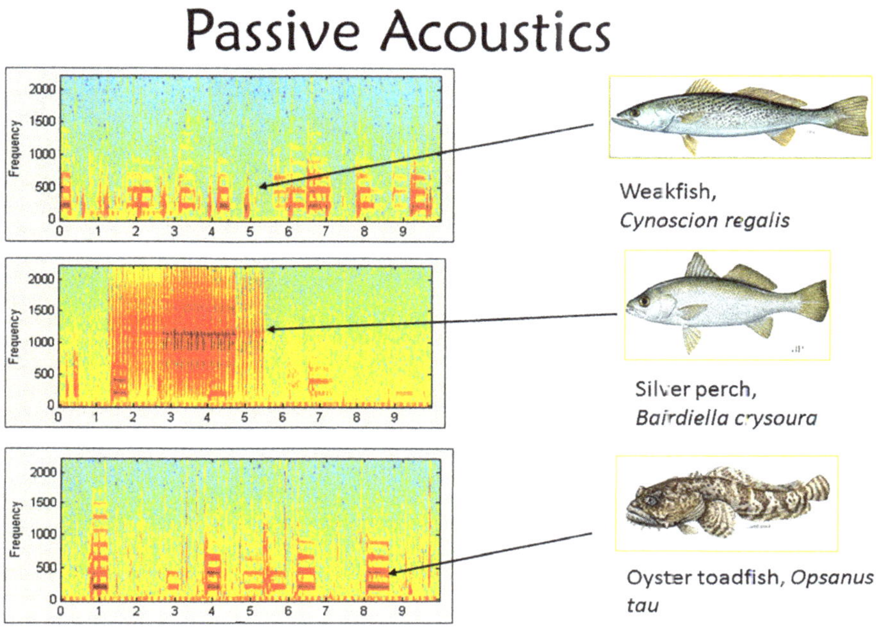

Figure 6.7 Sonograms that allow specific identification of fish sounds during courtship.

Temporal Scales

The dynamic nature of estuaries, like those surrounding the Sheepshead Meadows and adjacent islands, influences the flora and fauna and even occurrences of habitats at several scales or time frames. To capture this variation, we sample in all seasons. Perhaps the most influential variable is the high and low tide frequency and duration, which can expose the bottom or leave it submerged for minutes to hours. This variable obviously influences the distribution of many aquatic animals such as fish, crabs, and shrimp, as we often observe while capturing animals with seines at various tidal stages. Other sources of variation include the differences between day and night, but also the transitions during dawn and dusk. At larger scales, the seasons and associated temperature changes, which are especially pronounced in temperate estuaries, exhibit great influence on the plants and animals which live in the Sheepshead Meadows area. At the scale of years, a different set of changes can occur in response to the dynamics of some populations and climate-based influence on our long-term sampling programs, including fish larvae, using a plankton net, juvenile fish and crabs, with trap sampling, and a wider variety of species, sometimes including

Figure 6.8 Undergraduate students from the Estuary in Winter field class studying the shellfish prey dropped by herring gulls on a dock at the Rutgers University Marine Field Station.

diamondback terrapins and large adult fish, using otter trawls. Other study area changes occur at sporadic and relatively unpredictable frequency such as storms, ranging from nor'easters to hurricanes, and upwelling events, when cold water from the ocean enters the estuary. In many cases, Rutgers University students participate in these sampling opportunities (Fig. 6.8).

Notes from the Road—A Short History of Long-Term Fish and Crab Sampling

We generally lack perspective on the changes occurring underwater and particularly for animals such as fish and crabs that vary in the type and number within the salt portions of the estuaries, including the marshes and waters comprising Sheepshead Meadows. With fortunate foresight we, at the Rutgers University Marine Field Station, began three long-term sampling programs, ongoing for decades, which provide important insights throughout the program. A need-to-know which fish spawned in the ocean prompted the initial study. These fish provide larvae that grow and survive within the estuaries and eventually contribute to the adult population that might become harvestable in commercial or recreational fisheries or that play an important role in coastal ecosystems. Our approach was to capture these small, recently hatched fish in fine mesh (1/25 inch) plankton nets suspended in mid-water. Knowing when the fry arrived from the ocean proved critical, so we sample on flood tides as the Atlantic Ocean waters enter the estuary through Little Egg Inlet and up Little Sheepshead Creek. We sampled at night, thereby assuring that the larvae could not see and avoid our nets, which validated a representative sampling. The bridge over Little Sheepshead Creek provided an ideal convenient location close to the Station. To ensure good coverage, we sampled frequently—once a week—and have done so for over three decades. To date, an estimated over 700,000 fish larvae underwent capture and identification.

With the same rationale, ongoing sampling using small "minnow" traps (1-inch opening with ¼-inch mesh) serve to capture slightly larger juvenile fish conveniently within our boat basin, just off Great Bay, at least twice a week, since 1992. These collections have produced fewer individuals (approximately one-quarter million), but they represent a critical period in the fish's life history. In addition, because we wanted to know how these animals use different habitats, we began sampling in 1992 with otter trawls (16-foot mouth opening, ¼-inch mesh) in different depths, in different representative habitats (marsh creeks, bays, channels, etc.) and standardized the locations in the Sheepshead Meadows vicinity. These samples, collected every July and September, have helped us accumulate information on thousands of juvenile and adult fish that other programs fail to sample. Together, these unique efforts have helped us describe the early life history of various species, including summer flounder, winter flounder, blue crabs, and bay anchovy. We also determine critical habitats for a variety of species, evaluate the year-to-year changes in the abundances of these and other species, and follow the changes in the fish fauna in response to warming water temperatures.

Experimental Approaches

In the Sheepshead Meadows we have often used a variety of experimental techniques, both in the field and the laboratory, to identify patterns. For example, we have caged small juvenile fish to gain very specific growth measurements within individual habitats, such as experiments with summer flounder (Fig. 6.3) and other species. In other instances, we have brought animals into the aquaria and manipulated temperature and day-length as an impetus for reproduction during the mid-winter period, when they typically only spawn in the summer. This allows us to extend our research season. In other laboratory experiments, we have focused on habitat choice experiments in large tanks with flow-through seawater from the adjacent estuary, both during the day and night. Other observations have emphasized tag evaluation, including their retention on fish of a variety of sizes.

In another approach, Rutgers eDNA Lab researchers are conducting environmental DNA (eDNA) sampling from the Little Sheepshead Creek bridge and in a marsh pool near the Rutgers University Marine Field Station as a method for characterizing the fish diversity in these habitats. These researchers collect water samples during a range of tidal stages and then filter these samples to concentrate the remaining DNA in the water column from living organisms. Researchers then extract the DNA from the filters, amplifying the fish DNA present in each sample, and sequence the resulting DNA. The completed sequences are matched to a genetic reference library to identify and generate a list of fish species from each water sample. Another research goal is to evaluate eDNA to detect fish diversity through a comparison of those detected using eDNA sampling to those collected at this site during the long-term plankton net sampling. Environmental DNA is an incredibly sensitive tool for detecting rare species, making it possible to determine if fish are shifting their range due to climate change.

The eDNA dataset will provide this evidence prior to being physically caught in a net. Yet another goal is to better understand the spatial and temporal scale of the eDNA "signal" present in the estuary.

The Rutgers University Marine Field Station is also conducting an ongoing study of nor'easters beyond the observational record as an index of climate change, thereby providing a modern analog. The approach utilizes eDNA as a proxy to identify these storms and their modern overwash deposits, determining the presence of marine eDNA. The eDNA has the potential to distinguish nor'easter overwash deposits from other high energy events due to microbial community differences within the marine environment during such an event. The ultimate aim is to apply this methodology to create a chronological composition of nor'easter events from the sediment record over the past 2000 years.

Yet another intriguing research project in the relatively unaltered salt marshes along Seven Bridges Road at the Rutgers University Marine Field Station is an attempt to determine seasonal, annual, and longer-term variation in the foraminifera composition, single celled organisms with hard shells (tests), that live in marsh sediments. These tests, when removed and identified in salt marsh cores, prove useful because of their relationship with sea level over time.

Another level of insight can be determined at a much smaller scale, even under a microscope, of select animal tissues. These can include analysis of stomach contents to determine feeding habits. Additional insights can be gained through examining fish parasites to understand the parasite's life history and the impact they may have on the host fish—like black sea bass. Researchers have used the same general approach through examining the scat (fecal material) of harbor seals and identifying the strikingly different ear bones (otoliths) of individual fish species (Fig. 6.9) to determine the fishy parts these animals consume when they are present in the study area during the winter. Further, the examinations of fish otoliths (ear bones) allow for aging and back calculating their life history from years to days. These can also include analyses of specific elements to determine a history of habitat use with microchemistry techniques, even for larval fish. Together these multiple approaches, at different temporal and spatial scales, provide insights from this relatively undisturbed area.

Figure 6.9 (Following page) Variation in relative size and shape of representative fish otoliths from different species that are used in a variety of studies from age and growth to analysis of stomach contents of predators. Prepared by Paolo Lopez-Duarte.

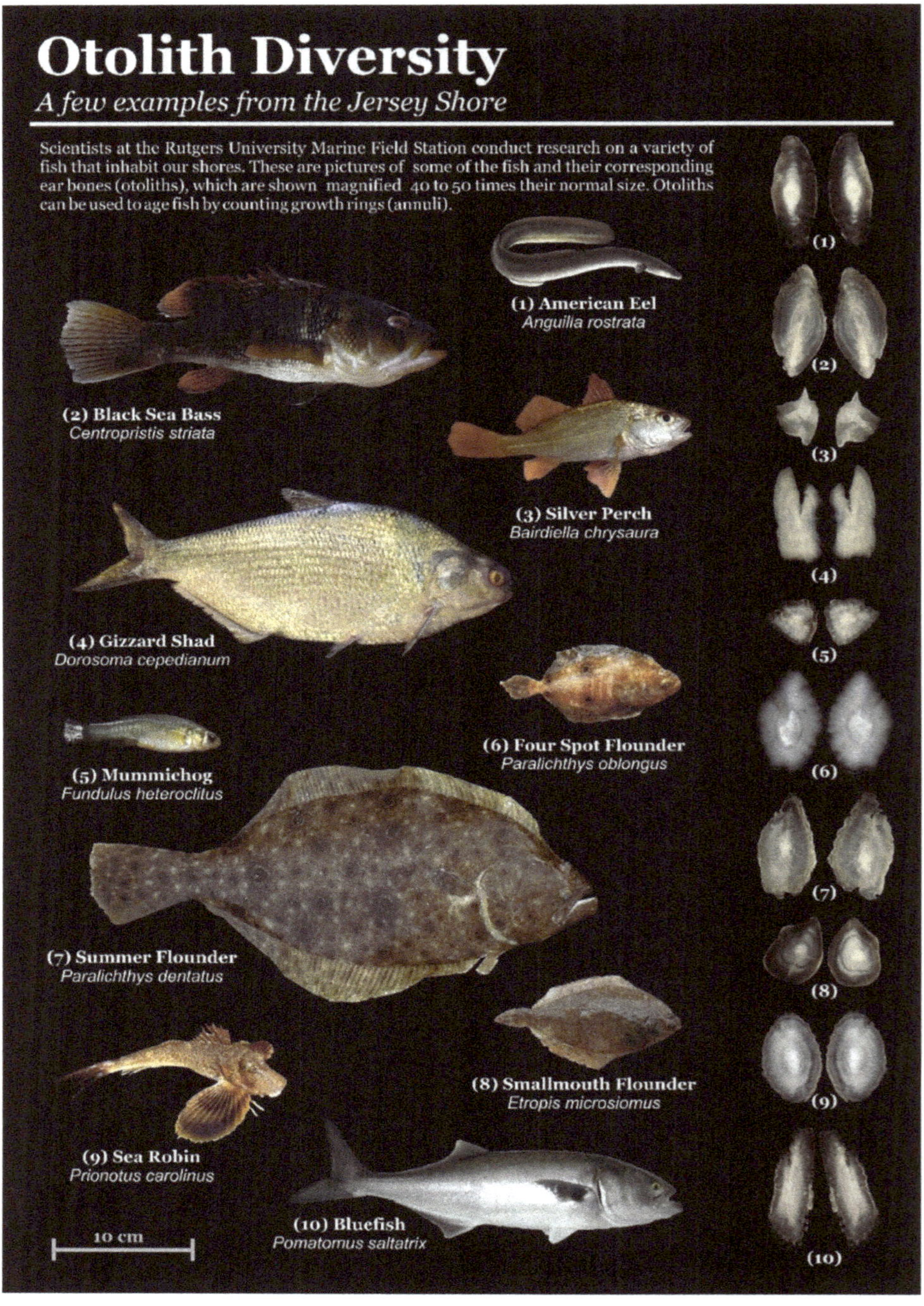
Otolith Diversity
A few examples from the Jersey Shore
Scientists at the Rutgers University Marine Field Station conduct research on a variety of fish that inhabit our shores. These are pictures of some of the fish and their corresponding ear bones (otoliths), which are shown magnified 40 to 50 times their normal size. Otoliths can be used to age fish by counting growth rings (annuli).
(1) American Eel
Anguilia rostrata
(2) Black Sea Bass
Centropristis striata
(3) Silver Perch
Bairdiella chrysaura
(4) Gizzard Shad
Dorosoma cepedianum
(5) Mummichog
Fundulus heteroclitus
(6) Four Spot Flounder
Paralichthys oblongus
(7) Summer Flounder
Paralichthys dentatus
(8) Smallmouth Flounder
Etropis microsiomus
(9) Sea Robin
Prionotus carolinus
(10) Bluefish
Pomatomus saltatrix
10 cm
(1)
(2)
(3)
(4)
(5)
(6)
(7)
(8)
(9)
(10)

Chapter 7
Habitats

The habitats in the Sheepshead Meadows and adjacent islands and Little Egg Inlet comprise natural and human-altered sites that transitioned from their natural state (Fig. 7.1). In many instances these transitory changes provide the ability to understand what is occurring in estuarine watersheds, with the natural habitats providing important baselines. Many human-altered habitats retain the structures and function of natural habitats, a relevant consideration. Further, we need to understand the varying ecological role and significance these different habitats play between the seasons in this temperate area.

Natural Habitats

The Sheepshead Meadows, the adjacent islands, and Little Egg Inlet largely consist of relatively unaltered natural habitats. These areas comprise salt marshes, creeks, bays, and thorofares. Each natural feature provides important ecosystem services (Table 7.1), especially in their natural state, although their role can vary with seasons, tides, and day vs. night.

Little Egg Inlet

With Little Egg Inlet's proximity to the Sheepshead Meadows and Seven Bridges Road (Fig. 1.1), the inlet has a great influence on all these habitats. This occurs because the inlet serves as a conduit between the ocean and the estuary which, in turn, influences almost all environmental characteristics including salinity, temperature, dissolved oxygen, and pH. The inlet also provides transport of larva and juveniles of fish, crabs, shrimp, and plant production and nutrients to estuarine habitats. The same inlet offers a seasonal migration route to overwintering areas along the Jersey coast and along the entire East Coast for the individual aquatic fauna listed above. Little Egg inlet is one of the few natural New Jersey inlets, which further enhances its connectivity. With the Forsythe National Refuge protecting both sides of Little Egg Inlet—Holgate Wilderness Area to the north and

Table 7.1. Ecological and socioeconomic services that natural, parallel grid ditched, and Open Marsh Water Management (OMWM) marshes provide; table modified from Peterson et al. 2008 and is based on our own research and observations. A "+" indicates services available, "-" indicates services unavailable, and "?" indicates unknown.

Services	Natural Marshes	Parallel Grid Ditched Marshes	Modified OMWM Marshes	Marsh Platform Under Sea Level Rise
High production at food chain base	+	+	+	?
Small fish and crustacean refuge and foraging grounds	+	+	+	?
Larger crab and fish feeding grounds during high water	+	?	?	?
Wildlife habitat	+	?	+	?
Buffer against storm-wave damage	+	+	-	-
Shoreline stabilization	+	+	?	?
Hydrologic processing	+	+	?	?
Water quality	+	?	+	?
Biodiversity preservation	+	-	+	?
Carbon storage	+	+	+	?
Aesthetics	+	-	-	-
Natural Heritage	+	-	-	-
Ecotourism	+	+	+	?
Education	+	+	+	?

Little Beach to the south—this protection renders the inlet effectively natural with minimal dredging. These characteristics all serve to enhance habitat and faunal diversity behind the inlet and the adjacent ocean.

Salt Marshes

Marshes comprise various sub-habitats, including the vegetated marsh surface, pools or ponds, intertidal creeks, subtidal creeks, and fringing marsh (Fig. 7.1), which together can provide varied ecological services. Whether it is a natural marsh, a marsh modified for mosquito control with parallel grid ditches, or Open Marsh Water Management, they can function similarly (Table 7.1).

Figure 7.1 Aerial photograph at the end of Seven Bridges Road where it meets Shooting Thorofare and sand washovers are created during storms and over time as has occurred adjacent to the Rutgers University Marine Field Station at the end of the causeway leading to Seven Bridges Road. Erosion of the road and redistribution of the original sand and gravel is evident at the end of the road. The former fish factory on Crab Island (upper right) is evident on the islands formed as flood tidal deltas. Numerous meandering tidal creeks are common on the marsh surface. These include Willit Creek adjacent to the Marine Field Station and Hatfield Creek between the Station and the fish factory.

Marsh Edges

The salt marsh edges, including those on bays, creeks, thorofares, and ponds, provide dynamic interactions. These interactions with the adjacent water vary, however, based on storms, tide range, and wave fetch, including the water's atmospheric interactions at low tide. In other instances, the fauna either influences or is influenced by the marsh edge. In addition, the water-to-marsh edges respond dynamically to a number of meteorological and ecological events and activities.

Storm-generated overwash fans create a striking dynamic change, altering the marsh edges over time, as witnessed in a series of aerial photographs (Fig. 7.2). In these instances, the wave and current action drop sediments on the marsh surface, which buries the vegetation, but also elevates the substrate. At the time of this writing in 2023, the sand deposition has reached the Rutgers University Marine Field Station boat basin. Over time, the sand has completely covered the vegetation over an ever-increasing territory, and the sand depth has continued to increase (Fig. 7.3). During the early stages of overwash, the new sediments may assume other functions, including as nesting area for diamondback terrapins. Vegetation may grow up through the new sediments over time, depending on depth, or colonize the sediments from seedlings and establish another, higher marsh surface. This process is a general one, based on aerial photographs, historic charts, and vibracores collected at nearby Brigantine that demonstrate the source of these storm-generated changes over decades and centuries.

The marsh crab and its herbivorous activities present another process that alters the water-to-marsh edge. They consume the saltmarsh cordgrass's aboveground leaves and belowground roots and rhizomes, and associated burrowing activity causes a salt marsh die-off, leaving the marsh edge eroded and unvegetated (Fig. 7.4). We derived evidence for this process within the Sheepshead Meadows from drone observations and walking the marsh surface, particularly in Hatfield Creek (Fig. 7.1). The result of such nocturnal activity is most obvious along creek edges between the high and low tide levels. Fiddler crab burrows are also evident along marsh edges throughout the marshes, especially along creeks. Intertidal marsh peat calving and its subsequent transport into subtidal areas provide yet another dramatic marsh-edge altering process (Fig. 7.5). We have determined the distribution and abundance of these peat reefs with visual and sidescan observations in the Sheepshead Meadows thorofares and on the adjacent

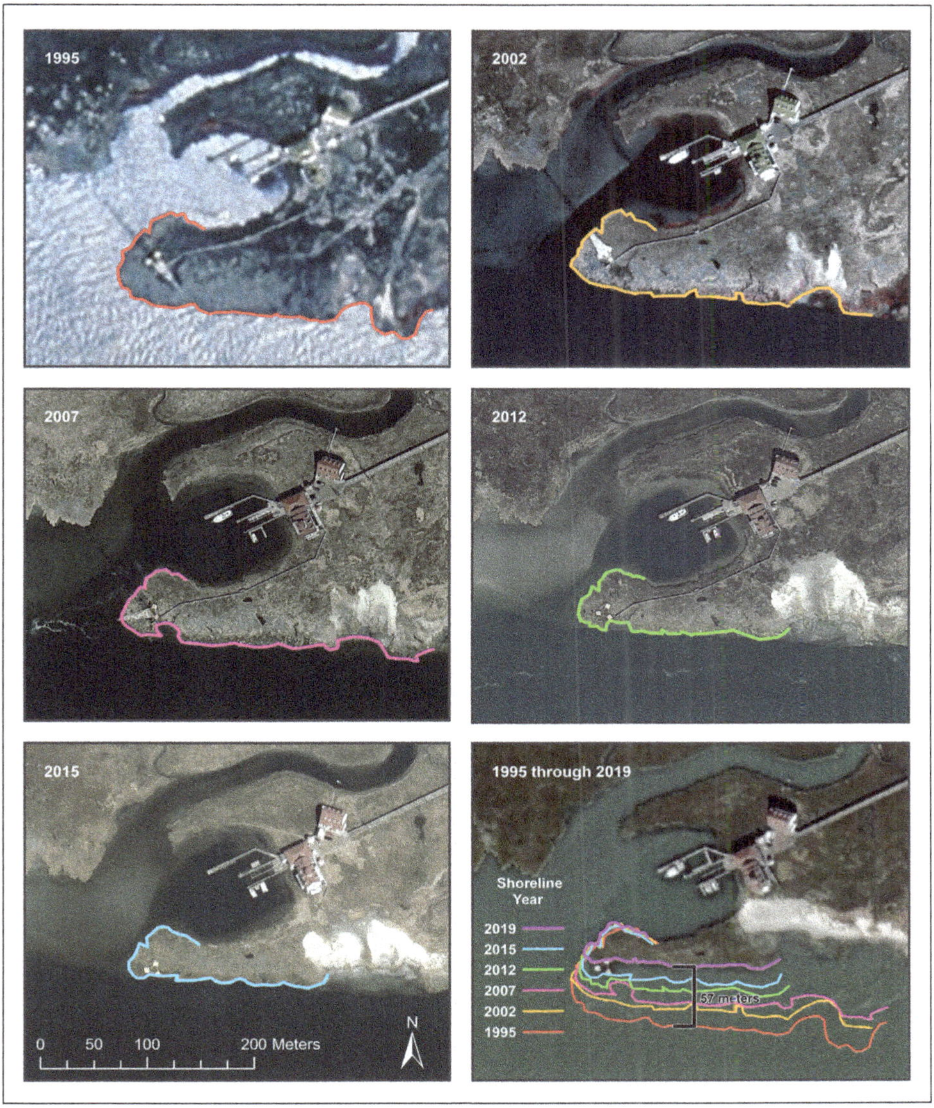

Figure 7.2 Shoreline erosion at the edge of Shooting Thorofare adjacent to Rutgers University Marine Field Station boat basin from 1995 to 2019 based on aerial imagery. Note also the extensive areas of overwash sands that are evident at the marsh edge in the lower right of each photo but with increasing extent over time.

Figures 7.3–7.4 (Top) Saltmarsh being covered by moving sand which covers the marsh grass but also serves to elevate the marsh surface. (Bottom) Unvegetated edge as the result of feeding on marsh grass by burrowing marsh crabs at Hatfield Creek.

Figure 7.5 Portion of salt marsh peat calving off the marsh edge into a thorofare during the winter in the Sheepshead Meadows.

islands. In general, they occur most commonly on the outer, deeper edges but their occurrence varies between individual thorofares.

The point where saltmarsh cordgrass and salt hay vegetation types converge provides another prominent habitat edge (Fig. 7.6). The two plants differ in structure: salt hay vegetation has a deep thatch layer (living and dead stems and leaves between the growing vegetation and the soil), while saltmarsh cordgrass features a much sparser thatch layer. The vegetation edges themselves differ in that they can have less litter accumulation, faster litter decomposition rates, a warmer and drier microclimate, and lower elevations. Wrack provides another prominent habitat (Fig. 7.7). A common natural disturbance, variable in time and space, wrack comprises dead marsh vegetation which waves and tides deposit onto the marsh. It can physically disturb the marsh by damaging live vegetation, but it can also increase nutrients, organic matter content, and create new microhabitats.

Marsh surface pools are dynamic over time based on inorganic and organic sediment accumulation and sea level rise. Pooling on the Sheepshead Meadows vegetated marsh surface (Fig. 7.1) and the adjacent islands

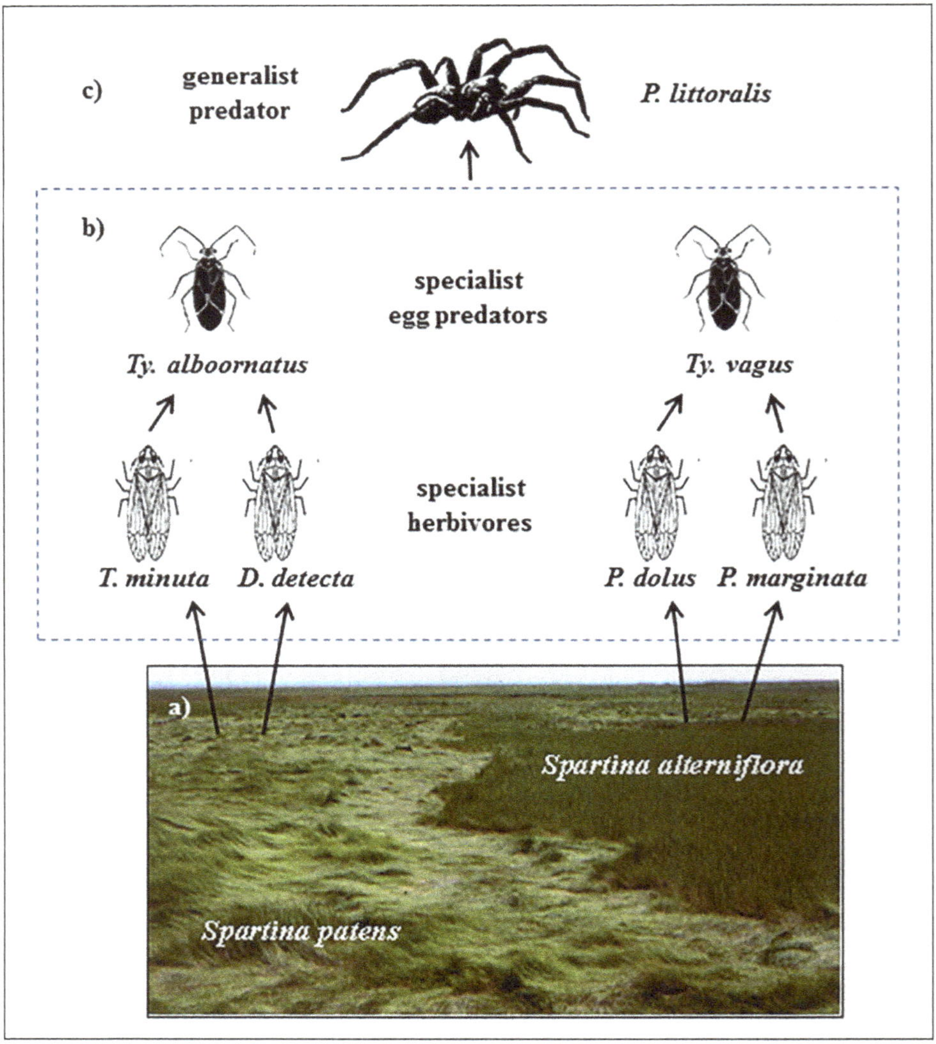

Figure 7.6 Edge between saltmarsh cordgrass and salt hay along Seven Briges Road with associated fauna of insects and spiders. From Wimp et al. (2011).

vary from flooded depressions to different-size, natural marsh pools as well as those that grow with sea level rise, referred to as "lakes." A small subset of marsh pools near the Rutgers University Marine Field Station range in depth from 1–1.5 feet and 5–60 square yards though some measured up to 143 square yards. Different vegetation create major differences between pools. Deeper pool bottoms usually feature dense stands of widgeon grass. Other marsh surface pools can be covered with numerous types of algae

Figures 7.7–7.8 (Left) Dead marsh vegetation or wrack (saltmarsh cordgrass, common reed) lays over the marsh surface (center and lower left). Linear marsh pools typically parallel Seven Bridges Road especially at the southeastern end. Note tall, dead common reed from the previous year along the road with telephone poles sticking above them (upper left) with new, shorter shoots just coming up in this image from May 2021. (Right) Marsh pool with algae (light green) floating at the surface surrounded by patches of saltmarsh grass (darker green) as typically occurs during the summer. Drone image by Rob Auermuller.

during the summer (Fig. 7.8). The coverage can vary from complete to only pockets of algae at the surface, depending on location, temperature, and perhaps other variables. The algae disappear during the winter.

The pools' environmental conditions can vary dramatically over long and short periods. Pool water salinity can vary seasonally and with the amount of rainfall. During the summer, when temperatures rise, evaporation may increase the salinity up to and over 40 ppt. The summer reading of the water's dissolved oxygen can vary from night to day. During the day, solar radiation (sunlight) warms the pool water temperature, causing bacteriological photosynthesis which supersaturates the pools' oxygen level. At night, however, when photosynthesis ceases, the same bacteria sucks the oxygen out of the water. The oxygen level can decline to zero only to rebound again as the sun comes up. During the winter, solar radiation and temperature influences remain obvious (Fig. 7.9), but not as extreme for dissolved oxygen as in the summer heat. In some years, the ice-covered pools combined with low water temperatures can cause winter fish mortality. Marsh pools frequently change water color, presumably due to algal/bacterial composition changes over time. Frequent photographs taken along the causeway leading to the Rutgers University Marine Field Station provide evidence for this change (Fig. 7.10). During the summer of 2018,

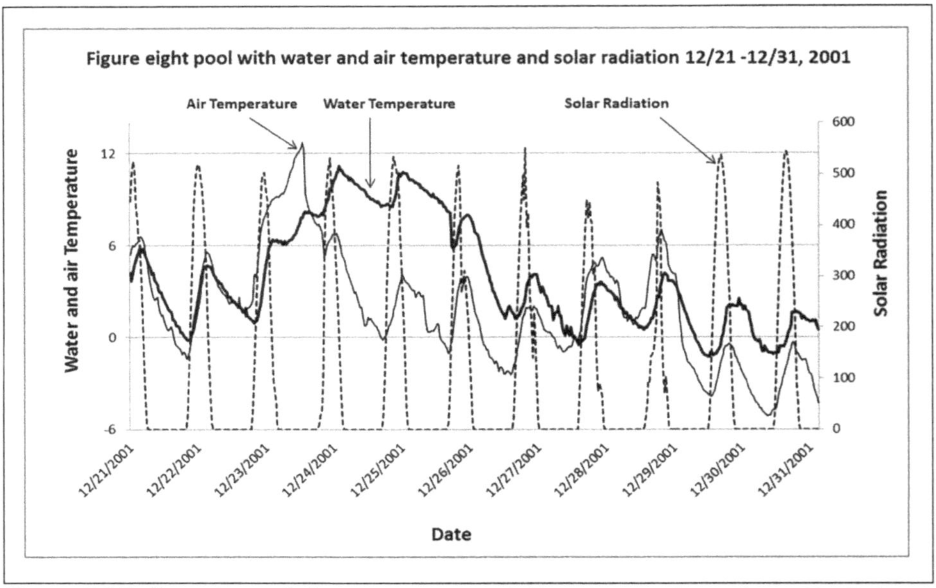

Figure 7.9 Daily water and air temperature variation at a marsh pool near the Rutgers University Marine Field Station as measured with data loggers indicating the obvious influence of solar radiation on air and water temperature during December 2001. Prepared by Uwe Kils.

the marsh pools displayed multiple colors, a common occurrence for days to weeks at a time. The marsh's smaller, shallower surface depressions also often serve as nursery habitat for some species. These include the recently hatched mummichog of 0.5–1.5 inches, and all stages of the smaller, cryptic species, the spotfin killifish.

Salt Marsh Waterways

The Sheepshead Meadows feature various waterway types. Some waterways exchange water with marshes and adjacent bays during tidal circulation and storms. Dead-end intertidal and subtidal creeks exchange water on flood tides and ebb tides. Thorofares pass water through marshes to and from adjacent bays. Many intertidal creeks' upper limits have mudflat endings surrounded with marshes. Low tides provide the evidence, exposing the mud, and the small creek channel remnants are clearly visible (Fig. 7.11). Often these creek ends feature small grass islands interspersed within the mudflat, perhaps representing lower sea level remnants. Small rivulets often fracture these mudflats and penetrate the marsh.

Figure 7.10 Composite photographs of a single marsh pool located near the gate to the causeway to Rutgers University Marine Field Station (upper left) with variation in color over time including clear water (upper right, 6 May 2019); bright green (middle left, 11 July 2018); milky yellow (middle right, 4 July 2018); white/green with algae around the perimeter (lower left, 22 July 2019); and the same pool two days later (brownish yellow, 24 July 2019).

Figure 7.11 Upper limit of an intertidal creek in unaltered marsh at the southern end of the Sheepshead Meadows at low tide with exposed mud and patchy salt marsh vegetation. Drone image by Rob Auermuller.

Willit Creek, a small (approximately 3000 feet long), meandering dead-end creek serves as a subtidal creek example. Sampled frequently, Willit Creek possesses characteristics typical to many Sheepshead Meadows natural subtidal creeks. The water depth, though very irregular, ranges from 2–10 feet and experiences numerous abrupt depth changes (Fig. 7.12). Creek bends often feature deep holes. While Sheepshead Meadows' many creeks exhibit these traits, like New Creek, for example, the adjacent Foxboro Creek features different traits (Fig. 7.12). Bottom sediment grain size varies along Willit Creek, as a representative example, with low percent silt/clay sediments near the mouth, high percent silt/clay sediments in the upper portion dead-end, and medium grain size in between (Fig. 7.13). Similar water column characteristics in temperature and salinity registered across the range of creek sites sampled (Fig. 7.14). However, there was a distinct, tide-related difference in dissolved oxygen along the creek with similar values at the lower station across all stages of the tide but distinctly lower levels on night high and low tides at the upper station. The values at the intermediate station reached their lowest levels on the daytime low tide.

Multiple thorofares (or thoroughfares as used by some authors and maps) flow through the saltmarsh cordgrass of the Sheepshead Meadows peninsula and adjacent islands. Five thorofares, each bridged on Seven

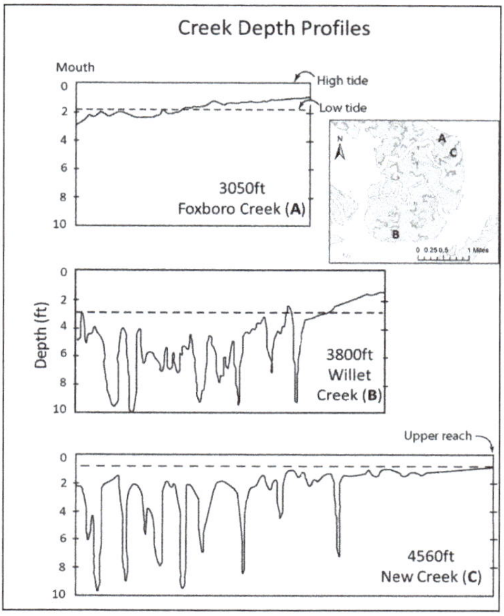

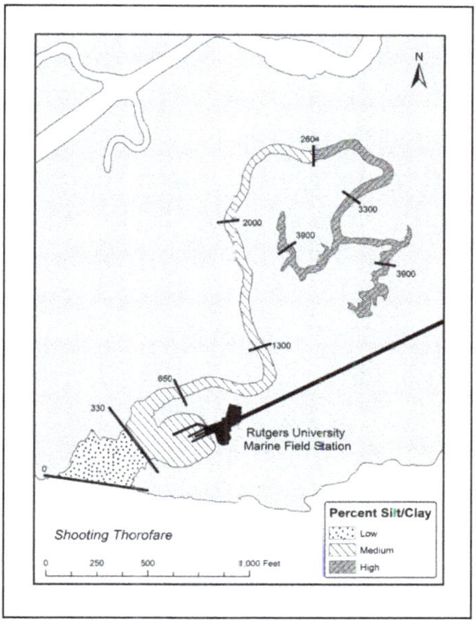

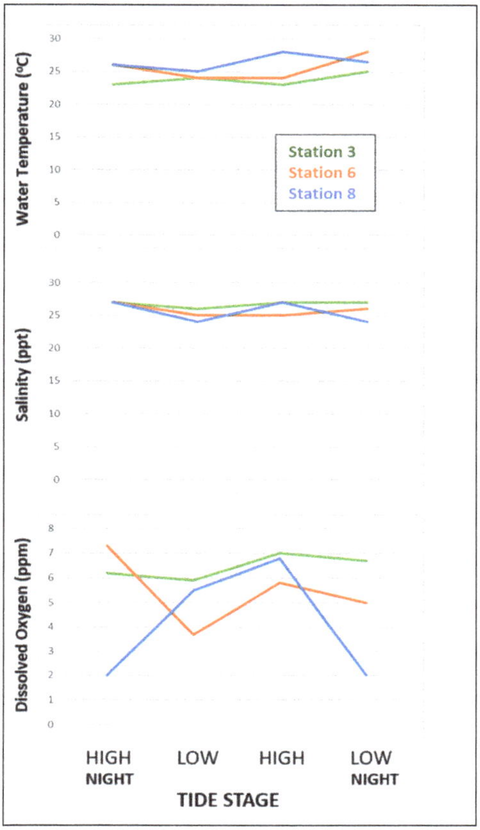

Figures 7.12–14 (Top left) Water depth in subtidal Willit Creek (middle) and New Creek (lower) is very variable throughout most of the creek with some holes ranging to 10 feet. Subtidal Foxboro Creek (upper) depth is more uniform. See inset for locations. (Top right) Sediment characteristics along the long axis of Willit Creek. The creek is subtidal for much of its length. (Bottom left) Water column properties relative to high and low tides during the day and night in Willit Creek during July 1990. Numbers indicate specific station locations from lower to upper creek. See inset for location of creek. Figures prepared by author and Ryan Larum.

Bridges Road, cross the Sheepshead Meadows' main axis (see Fig. 1.1). Each of these thorofares vary in depth and other topographic aspects but feature similar salinity (24–32 ppt), with those furthest from the ocean waters and Little Egg Inlet possessing slightly lower salinity. The depths are different within and between thorofares. Little Sheepshead Creek contains the deepest depth, measuring 9–18 feet, while Little Thorofare's depth measures 3–6 feet. Big Thorofare depth measures 3–16 feet, while Big Sheepshead and Jimmies Creek are the shallowest at 6–9 feet and 3–9 feet, respectively. Curiously, the tidal flows vary between them as does the circulation from the open water bays that surround the meadows. The high tidal flow is from Little Egg Harbor to Great Bay in Little Sheepshead Creek and from Great Bay to Little Egg Harbor in Big Thorofare, Little Thorofare, Jimmies Creek and Big Sheepshead Creek. There are also seasonal temperatures differences and degree of ice cover, at least in cold winters. The greater depths and warmer temperatures in Little Sheepshead Creek make ice cover less frequent, presumably due to the warmer oceanic temperatures. Sidescan sonar data demonstrate that the thorofares through the peninsula contain peat reefs (Fig. 7.5), ranging from less than one-half a foot to 49 feet in length, with Little Sheepshead Creek and Little Thorofare possessing the greatest peat reef density. The thorofares between the Seven Islands and Story Island complexes (see Fig. 1.1, 1.2) share many characteristics with those through the peninsula including temperatures, salinities and depths, as well as the variability in peat reef distribution.

Flood Tidal Delta Islands

Two island complexes called flood tidal delta islands flank both sides of the Sheepshead Meadows. These complexes formed from flood tides carrying ocean sediments (see Fig. 1.1, 1.2 and Chapter 2). For those that take an interest in these islands, be aware that the names have changed over time. Recent maps, prior research, in combination with drone imagery, aid our understanding of these low-lying islands and their structure. Based on 50-year studies of older, Great Egg Harbor flood-tide delta islands, we know that they are losing area (less than 5%), and that their orientation has shifted to the northeast, although there is individual variation.

These centuries old island features dominant Great Bay and Little Egg Harbor, yet little knowledge exists concerning these features. Intertidal marsh grass dominates these islands as on Sheepshead Meadows. They provide excellent waterfowl habitat, based on the gunning clubs that occupy the islands. Nesting gulls find important habitat here, too and herons and egrets feed here, along with numerous shorebirds on migratory pathways. Underwater, the islands' numerous thorofares provide abundant and productive fish habitat, based on the recreational fisheries for weakfish, bluefish, summer flounder, and other species during spring through fall and commercial fisheries for blue crabs. These sandy to muddy channels also serve as hard clam habitat.

In the Story Island complex (Fig. 7.15), the largest island lends its name to this complex and forms the northern boundary for Marshelder Channel and another deep channel at Soldiers Hole, which in turn is bound by Goodluck Sedge. At the other extreme, Long Beach Island forms the boundary between the adjacent estuary and the Intracoastal Waterway. Marshelder Channel, at the western end of this complex, contains Parker Island as the channel proceeds into Tuckerton Creek. Parker Island is now much smaller. Adjacent to them are Drag (formerly Hither Island) and Hither (formerly Goodluck Islands) islands on either side of Hither Channel (Fig. 7.16). At the complex's eastern side, the larger Barrel Island dominates a sizeable portion of the Intracoastal Waterway's western edge (Fig. 7.17).

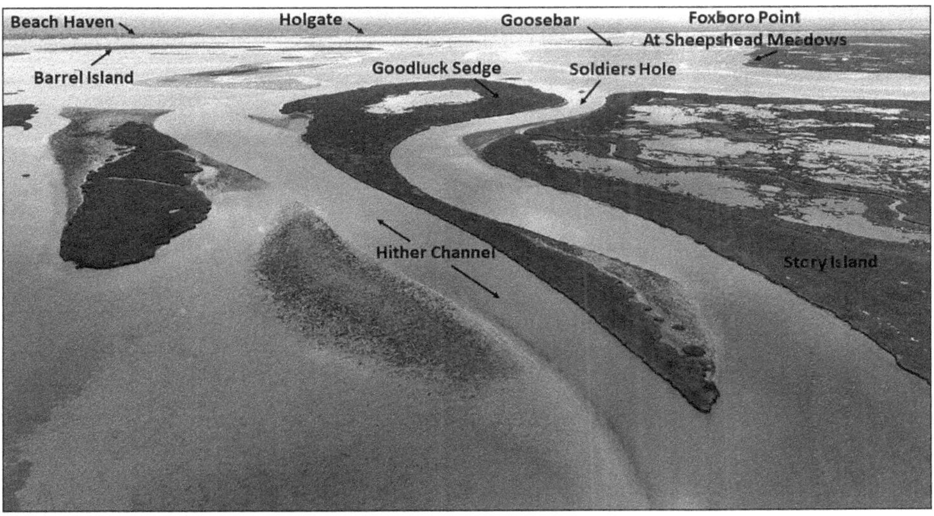

Figure 7.15 Image at the southeastern end of the Story Island complex of islands west of the southern end of Long Beach Island (upper) and north of the Sheepshead Meadows at Foxboro Point (upper right). The eastern end of Story Island, with large numbers of large marsh pools, is evident (center right) while the sinuous Goodluck Sedge is in the center of the image. Some margins of the islands and channels have vegetated intertidal flats as the one adjacent to the Hither Channel label. Drone image by Rob Auermuller.

Figures 7.16–17 (Top) Two of the larger islands in the Story Island complex are dominated by Hither Island, which includes large numbers of marsh surface pools (upper), and Drag Island (lower) that are separated by deeper channels through shallow, sandy flats. Two gun clubs are evident on Hither Island while a single one is evident on Drag Island based on this drone image from December 2021. (Bottom) Barrel Island is one of the largest islands at the eastern end of the Story Island complex. It is bordered on the east (left) by the Intracoastal Waterway and the adjacent southern end of Long Beach Island at Beach Haven. The gray intertidal areas (center right, lower center) are vegetated, intertidal flats. Drone images by Rob Auermuller.

In between is Middle (formerly Johnny Sedge) Island. Numerous small to large non-tidal pools and tidal creeks dominate the surface on some islands.

Seven Islands feature fewer isles (even fewer than seven now) with fewer thorofares (Fig. 1.2). Crab Island comprises the largest isle, the former home of an Atlantic menhaden processing factory. Newmans Thorofare separates Crab Island from the Sheepshead Meadows. Little Crab Island also once featured an even earlier menhaden factory. Sloop and Pole thorofares separate the isles east and south of Crab Island. The Intracoastal Waterway lies east of Fish Island and also separates it from the smaller Tow Island. These island complexes both have extensive saltmarshes, with many marsh pools and creeks and thorofares dividing up the islands.

Sandy Beaches

The Sheepshead Meadows and the adjacent islands feature relatively few beaches. Extensive tidal action and/or high wave fetch that help to sort and deposit sand influence these typically low relief features. One of the easiest to observe is the one that extends from east of First Bridge (Fig. 7.18). While sand predominates, macroalgae from the adjacent bay covers it periodically, especially on very high tides and storms. Another prominent sandy beach lies at the end of Seven Bridges Road, where it faces Shooting Thorofare. Here, it is exposed to the thorofare's strong currents, as well as exposure from Little Egg Inlet, especially during storms (Fig. 7.1). This

Figure 7.18 Sandy beach bordering Little Egg Harbor along Seven Bridges Road adjacent to the bridge over Big Thorofare. In this low tide image large amounts of sea lettuce macroalgae are visible.

dynamic activity at this unprotected location results in an extensive sandy intertidal edge (Fig. 7.2). One reasonable interpretation is that some of the sand and gravel found in this edge originated and redeposited from the road construction beginning in the 1930s. Macroalgae and wrack frequently covers the edge, especially after high water periods. Another remote sandy beach exists at Barrel Island, part of the Story Island complex.

Tidal Flats

These typically sandy, intertidal features only appear at low tides. Worm tube colonies, other invertebrates, and their predators, such as blue crabs and attached macroalgae, all come into view at low tide along the edge. These tidal flats occur along the Sheepshead Meadows edges primarily in the portion nearer the inlet like Little and Big Sheepshead creeks. They also occur on portions of the intertidal areas near Seven and Story Island (Fig. 7.17) and Crab Island.

Bays

Great Bay and Little Egg Harbor Bay surround the Sheepshead Meadows and adjacent islands (Fig. 1.1). All of these bays provide access to the thorofares that flow through the Sheepshead Meadows. These waterways contain relatively clean water due to the low human population density and lack of industry, a continual pattern since natural resource industries declined 100–150 years ago. As a result, It is worth noting that these clean waters never experience low dissolved oxygen levels.

The bays' open waters typically share high salinity rates from Little Egg Inlet and the ocean; the relatively small freshwater amounts flowing from the Mullica River (into Great Bay) and Tuckerton Creek (into Little Egg Harbor) bear little influence. As an example, the waters near the Rutgers University Marine Field Station carry a constant salinity level in the high 20s. Precipitation typically measures approximately 39–49 inches/yr and holds little salinity influence most years. Seasonal changes affect the bay's average water temperature, ranging from 39–44°F in winter and 68–75°F in summer. During periodic summer upwelling, the ocean floods high-density cold water into the bays' lower portions through Little Egg Inlet. The shallow waters mix reasonably well (Great Bay averages less than 6 feet, Little Egg Harbor 6–23 feet), even though the bay features microtidal activity (range about 3 feet). Near the inlet, both bays' bottom sediments primarily comprise sand with increasing silt and clay upstream from the inlet. Great

Bay lacks eelgrass but remains productive from large macroalgae amounts and the surrounding marshes. Macroalgae is dominated by sea lettuce but also includes Agardh's red weed and graceful red weed. During summer and fall, these can cover the bottom and are exported offshore through Little Egg Inlet during storms and some tides.

Altered Habitats

While the area of interest remains largely unaltered, including those mentioned above, human activities have modified specific habitats and locations over the last century.

The Road

Sheepshead Meadows endured a major human alteration with the Seven Bridges Road construction and the accompanying bridges (Fig. 1.1), beginning in 1929. The completed roadway fostered other developments such as marinas, enhanced hunting and fishing access, and eventually, further development of the Crab Island menhaden reduction plant. In the process, the road, and its 100-foot right-of-way, altered the former salt marsh habitats. In essence, the roadway's increased elevation (Fig. 7.19) extended mainland vegetation types into the marshes (Fig. 7.20). For example, red cedar trees can be found from the Tuckerton mainland, where they are most abundant and largest, extending along the road to the very end where it meets Shooting Thorofare (Fig. 7.21). Other, deciduous trees and shrubs (Chapter 8) share the same pattern. The five elevated bridge approaches to the five bridges feature both flora types in greater abundance.

The roadway berm cut off numerous small rivulets that formerly meandered through the marsh. In some instances, some larger subtidal creeks continued flowing through multiple larger culverts installed under the road. In other instances, smaller culverts of varying number, allowed continuing flow like the creek on the approach to Capt. Mike's (Fig. 4.9). In more extreme situations, the berm completely cut the creeks, and water exchange could only occur during spring or storm tides. Over time, the road's end, where it meets the Shooting Thorofare waters, has eroded and the sand and gravel used during the original construction has been redeposited along the marsh edge (Fig. 7.1).

A thorough survey along the entire Seven Bridges Road also provided vegetation distribution information. The original survey, conducted 1973–1974, determined the potential effects from a power line that would

Figure 7.19 Light Detecting And Range (LIDAR) mapping imagery clearly depicts the difference in elevation between the higher (dark red) mainland and lagoon developments (upper left) of Mystic Islands in contrast to somewhat lower elevation of Seven Bridges Road. Other elevated portions within the Sheepshead Meadows include the fish factory on Crab Island (lower center) and dredge disposal sites on Story Island, on the westernmost portion of the road near the lagoon developments and at the mouth of Little Sheepshead Creek (lower right). There is some evidence of the higher elevations at each bridge where they cross the thorofares. Faint outlines of three confined disposal facilities can be detected on the right side at the upper end of the road. Prepared by Ryan Larum.

Figures 7.20–21 (Top) Portion of Seven Bridges Road including the First Bridge and its vegetated approaches, dominated by eastern red cedar trees, over Big Thorofare. Image by Rob Auermuller, October 2021. (Bottom) An image in winter from a helicopter shows the extensive marshes along both sides of Seven Bridges Road as it heads from the bridge over Big Sheepshead Creek toward the beach at Shooting Thorofare near Little Egg Inlet (lower left).

Figure 7.22 The side of the Grassle Pocket Marsh, named after Fred and Judy Grassle (pictured), that borders Seven Bridges Road near Oak Lane.

run from the then proposed Atlantic Generating Station in the ocean outside of Little Egg Inlet, and down the roadway's length to a power relay station near the other end (Fig 1.1). The survey extended from Willis Creek, about 1.2 miles south of Tuckerton, to the road's end at Shooting Thorofare. This examination included the paved road and its 100-foot right-of-way. Terrestrial vegetation occurs along the road while passing through the salt marsh center, because the roadway's berm's original construction featured a sand-filled bed (Chapter 4). This elevated roadbed also prevents saltwater inundation from the surrounding Great Bay and Little Egg Harbor. Fresh water from Willis Creek fed the area. When the survey occurred in the 1970s, a new lagoon development altered this small watershed, along with potential sea level rise. As a result, a dense salt marsh now dominates the area. The survey recorded 198 plant species, including marsh and terrestrial species. Of these, the road edge featured 52 introduced species. Coring trees during the survey produced an age report that listed groundsel trees as up to 10–12 years old. The other species date closer to the original road construction: eastern red cedar dated up to 31 years; bayberry up to 29 years; and black cherry up to 30 years, thus they colonized the elevated roadway soon after construction ended.

Common reed along Seven Bridges Road appears in several different settings. Along the immediate roadside from about the Grassle Pocket Marsh (Fig. 7.22, 7.23) to the road's end provides an obvious place for viewing common reed. The plant's occurrence, abundance, and height, however, can be quite variable. The plant does quite well at the roadsides' peak elevations and away from trees, reaching 10–12 feet (Fig. 7.24). At

Figure 7.23 Two of the sides of Grassle Pocket Marsh are bound by Seven Bridges Road (angling from the right) and Oak Lane (angling from the left) with the pocket marsh in between. The marsh is brown at the bottom of the picture (in Fall 2016, image from a helicopter). Tuckers Creek flows from the pocket marsh, through culverts under Seven Bridges Road and the creek entrance downstream to eventually join Tuckerton Creek. The pocket marsh is surrounded by maritime forest, up to Radio Road. Just off Seven Bridges Road within the forest in the lower right are the Annex to the Rutgers University Marine Field Station.

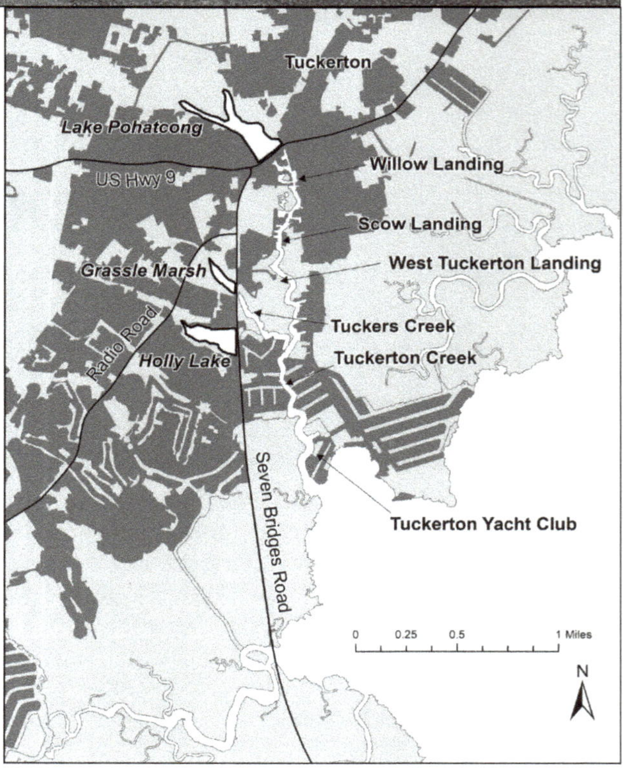

Figures 7.24–25 (Top) Saltmarsh grass growing right up to hardtop at lower elevation of Seven Bridges Road. At these lower elevations common reed is much shorter or completely absent but much taller at higher elevations as at the distant part of the road in this image. (Bottom) Historical locations along Tuckerton Creek from the dam at Lake Pohatcong to the mouth of the creek at Little Egg Harbor near Tuckerton Yacht Club. The locations of two man-made lakes (Lake Pohatcong, Holly Lake) are also shown.

lower elevations it is much shorter. Due to a reduced tolerance to saltwater, no common reed grows at the road's lowest elevations, which endure frequent saltwater flooding from the adjacent creeks and bays (Fig. 7.24). An upland forest comprising pine and oak trees surrounds the marsh with a few dead Atlantic white cedar in the watershed's upper reaches, appearing as a "ghost forest," reflecting the increasing salty water incursions due to sea level rise.

Pocket Marshes

Pocket marshes, i.e., small marshes, isolated from the main marsh and with diminished water flow, occur along Seven Bridges Road between the Big Thorofare bridge and Tuckerton. Seven Bridges Road construction and subsequent renovations since the 1930s resulted in these pocket marshes. Some water exchange still occurs through the culverts under the road (Fig. 4.9), a common approach in marshes and other natural habitats.

The least altered pocket marsh—Grassle Pocket Marsh—received its name from Judy and the late Fred Grassle in recognition of their contributions to marine and estuarine science (Fig. 7.22, 7.23). Water sources, both upstream and downstream, directly influence this watershed (Fig. 7.25). In this way, it shares many typical estuarine watershed characteristics, with freshwater coming downstream that mixes with saltwater coming upstream thus creating a salinity gradient. A small watershed releases freshwater, which flows downstream and into the marsh through culverts under Radio Road (formerly Island Road). Saltwater flows in with the tides from Tuckers Creek and enters the marsh downstream through culverts under Seven Bridges Road, which connect to Tuckerton Creek, and further downstream to Little Egg Harbor, the source of the higher salinity water. The marsh bordering the creek contains typical marsh plants including the dominant form, saltmarsh cordgrass, along the creek edge and in the higher salinity portion of the lower marsh. The invasive common reed dominates the marsh's upper portion, along the lower marsh's upland edge, and along the elevated Seven Bridges Road berm (Fig. 7.23). Aerial images taken over time provide insights into the Grassle Pocket Marsh invasion timeline. The earliest image from 1998 indicates common reed already present, but only in patches above the current boardwalk location and in the immediate boardwalk vicinity. By 2020 it had expanded dramatically so that it now dominates the upper marsh and even more of the lower marsh near Seven Bridges Road.

Figure 7.26 Linear ditches are common features of the northwestern portion of the Sheepshead Meadows. This dredged ditch joins seasonal, shallow, narrower mosquito control ditches on the marsh between Tuckerton Cove and Tuckerton Creek and lies behind Seven Bridges Road. The vegetation at this point varies from extensive red cedar trees along the highest elevation, with common reed at a slightly lower elevation and saltmarsh cordgrass lining both sides of the ditch.

Tuckers Creek's variable salinity influences the pocket marsh's flora and fauna composition, more so than anywhere else within Sheepshead Meadows. A submerged data logger and occasional salinity meter (refractometer) observations clearly indicate variable salinity ranging from 0–21 parts per thousand. The highest salinities occur at the lower end due to the downstream Tuckers Creek connection with Tuckerton Creek, Little Egg Harbor and the ocean through Little Egg Inlet. Tuckers Creek's upper portion typically possesses low salinity or fresh waters from streams flowing through two wooded areas that connect using culverts under Radio Road to the marsh.

The former Holly Lake Pocket Marsh, now the freshwater, non-tidal, shallow Holly Lake (Fig. 4.10) was once similar to the Grassle Marsh, an intertidal marsh connected directly to Tuckerton Creek (Fig. 7.25). With Seven Bridges Road completed, the Holly Lake Marsh connected to Tuckerton Creek through a culvert under the road. Historic images indicate water from Tuckerton Creek fed the intertidal marsh and creek, which is now within a lagoon development. At that time, a sinuous creek drained the marsh. Between 1956 and 1963, a newly constructed dam caused the marsh to flood, and it became the freshwater lake it is today, killing the original marsh vegetation. Aerial photographs detail the original creek channel draining the marsh after the spillway suffered damage from Superstorm Sandy and the lake drained. Today, some of the forest that surrounded the original pocket marsh still surrounds the lake.

Mosquito Control Habitat Alterations

Mosquitoes inhabiting salt marshes have likely plagued humans since the earliest indigenous peoples moved into what is now New Jersey. There are numerous accounts describing the impact of mosquitoes on the earliest European settlers and their livestock, especially among those who used the marshes for grazing their livestock, harvesting salt hay, and trapping along the tidal estuaries edges. Humans have made two major attempts to control salt marsh mosquitos through habitat alteration. The first attempt required cutting ditches through the marshes (Fig 7.26). This work began during the early 1900s and then was refined during the Great Depression's make-work projects.

The earliest mosquito ditching techniques generally comprise cutting parallel ditches across the marsh surface. In theory, the ditches drained the marsh, thereby preventing mosquito breeding in depressions on the marsh surface (Fig. 2.4). These ditches typically measuring less than three feet wide and two feet deep, appear in Sheepshead Meadows's northwestern portion on an early 1930s aerial photograph. Lagoon construction for Mystic Island and Osborne Island developments destroyed many mosquito ditches (Fig. 3.7). A study evaluated this ditching approach and its ecological significance through cutting new experimental marsh ditches near the First Bridge at Little Egg Harbor. These studies determined tidal flow influenced organic carbon movement with most carbon proportionally redistributed relative to creek depth, distance from the open estuary, and based on the precipitation amount. Creek types varied with generally less carbon exchanged in the mosquito ditches versus the natural creeks. By 1940, nearly all East Coast marshes featured ditching. One exception is the Sheepshead Meadows's southeastern portion and the adjacent islands.

Regular linear roadside ditches within the Sheepshead Meadows likely form part of a mosquito control drainage project (Fig. 7.26). Irregular ditches

Figure 7.27 Aerial image of Open Marsh Water Management (OMWM) habitats being constructed with a rotary ditcher. In this instance small ponds with small islands in the center are being created and connected to other ditches. The sediments from these activities are spread thinly over the marsh surface (left of the creek that bisects the image). All of this standing water provides habitat for fish that eat the mosquito larvae. Drone image by Ocean County Mosquito Control Commission.

frequently occur immediately adjacent and parallel to Seven Bridges Road, especially at the southeastern end (Fig. 7.7). We believe these form over time, as wrack vegetation accumulates along the road's higher elevations, which acts as a dam during storms. Once it is thick enough, it kills the normal marsh vegetation, prevents it from recovering, and facilitates the marsh peat rotting. This marsh portion eventually reduces its elevation and standing water becomes common except during a drought.

The newer Open Marsh Water Management (OMWM) approach creates a system of ponds and ditches on the marsh surface to link small, isolated mosquito-breeding depressions and tidal or non-tidal pools. These links allow larval mosquito predators, particularly fish, access for controlling larval populations (Fig. 7.27). Begun during the mid-1950s, the Sheepshead Meadows received this treatment during the 1960s. The pool and ditch shape and size vary because they are dug to connect individual mosquito breeding areas with fish habitats. These are distributed in the Sheepshead Meadows's western and central portions (Fig. 1.1). A detailed benchmark mapping effort for the meadows, based on 1990 and 1995 maps, determined that the mosquito-ditched portion contained fewer marsh pools relative to the OMWM-altered area and the non-ditched marsh portion.

Tuckerton Creek

A navigable waterway steeped in history, the Tuckerton Creek featured numerous marine-related industries along its banks, ranging from boatbuilding and shipbuilding to harvesting shellfish and finfish, as well as maritime commerce. The place names along the creek reflect these activities (Fig. 7.25) and appear in Tom Applegate's (2021) book containing photo comparisons. Willow Landing lies just below the dam, a location for mooring many sailboats, and eventually powered boats such as the SILVIA, a 65-foot-plus vessel, which once moved passengers, upwards of 80, between Tuckerton and Beach Haven on Long Beach Island. Nearby, Scow Landing served as the Van Sant shipyard location, which constructed boats measuring up to 100 feet. These landings provided a wintering location for the CYNTHIA, the Rutgers University research barge formerly towed to a variety of locations between Barnegat Bay and the Mullica River during the summer. Beginning in the 1920s, the Tuckerton Yacht Club became a well-known place to visit, located farther down the creek (Fig. 7.25). The yacht club burned down in 1973. Charcoal Landing stood just below the yacht club site, a place that provided a loading point for charcoal and lumber, harvested from the nearby extensive forests, on ships sailing to distant locations during the 1800s. Close by, Parsons Seafood, a five-generation business, has occupied the same location since 1909. Prior to the late 1920s, Green Street, which parallels Tuckerton Creek, did not reach the creek's mouth. The road was subsequently extended to its current location.

Today, the creek features the most altered habitat across the study area for varied reasons. Human development and adaptation within the creek watershed have resulted in significant modifications since Tuckerton's founding over 300 years ago. The creek currently stretches from the dam at Lake Hopatcong to its mouth at Little Egg Harbor (Fig. 5.22). The dam's construction is the most significant creek change because it influences water flow, water quality, and the flora and fauna farther downstream. In 2013, officials ordered a fish ladder added to the dam, which theoretically would allow anadromous fish, including river herring, to move above the dam and enter the watershed's upstream freshwater portions for spawning. My fall Ichthyology class, despite making frequent observations, failed to identify juvenile river herring in Hopatcong Lake over multiple years. Infrequent fish ladder viewing during the spring has never identified migrating adult river herring. The dam, however, does not appear to serve as a complete barrier to upstream movements for early American eel life stages. Abundant glass eels and elvers have been very abundant in consistent collections below the dam over several years. The same Ichthyology classes have collected these, and the older "yellow eels" of this species, in seine and electrofishing sampling from the Lake Hopatcong beach, and, at larger sizes, upstream in Mill Creek at Nugentown Road. One wonders whether these eels can grow, swim over the dam, travel downstream, and make their way to their Sargasso Sea spawning area.

Over the creek's length, the salinity ranges from 6–23 parts per thousand with the lowest salinity below the dam, and the highest at the creek's mouth. Numerous other sources influence the creek waters throughout its length, including land-based, impervious-surface runoff associated with roads, parking lots, houses, housing developments, businesses, and the Tuckerton Seaport. Marinas and extensive bulkheading have greatly modified the creek. A notable exception to the creek's cultural adaptation is the 17-acre wetland edge preserved by the Ocean County's Natural Land Trust off of Green Street downstream of the Dockside Restaurant.

Marinas and Boat Basins

The Garden State Parkway's construction during the 1950s provided enhanced access to the southern New Jersey coast on a regional level. On a local level, the on-again, off-again Seven Bridges Road construction provided earlier access (Chapter 3). Marina development, especially along the road, reflects that access. The marinas presumably developed soon after road construction began during the early 1930s. Newspaper articles appearing during the 1940s indicate that the road featured a number of rowboat, bait, restaurant, and other related businesses and they served both recreational and commercial fishers. The road also provided, at Little Sheepshead Creek, a landing location for boats transporting fishmeal and fertilizer from the Crab Island fish factory to Seven Bridges Road (Chapter 4). Since then, the basin's mouth has silted in and may now be closed at low tide.

Station 119, the U.S. Coast Guard facility, constructed during 1937–1938, concomitantly used its boat basin in carrying out its regular duties. Later, in 1972, Rutgers University purchased this abandoned facility and used the boat basin to support the Marine Field Station's research and teaching activities. Since the basin's construction it has become naturalized with fringing intertidal brown algae, green algae (both attached and free-floating), and a natural fauna comprising fish and crabs, based on twice-a-week trap sampling over 30 years. In addition, in recent years, oysters have colonized

Figure 7.28 Dredging of the Rutgers University Marine Field Station boat basin in September 2018. Drone image by Thomas Johnson.

pilings and the basin's marine railway. Dredging has occurred within the boat basin five times since its 1937 creation (Fig. 7.28). A detailed study in 2003 did not find any fish fauna change from before, during, and after the dredging.

In recent decades, several marinas have served the recreational and commercial fishing activities along Seven Bridges Road (Fig. 4.19, 4.20, 4.21), from those at the Big Thorofare bridge (formerly First Bridge Marina); Rand's Marina (abandoned after Superstorm Sandy, now open to the general public for launching boats and fishing/crabbing); Cape Horn Marina at Cape Horn (now also serving as a clam aquaculture operation); and Capt. Mike's at Big Sheepshead Creek. Besides the transition in ownership at Rand's Marina, and the removal of its building that once stood there, marsh grass recolonization has enhanced the site and features a "turtle garden" at one end of the parking lot. Sand added to the site (Fig. 7.29) successfully transformed the location into a diamondback terrapin nesting area as part of a habitat restoration project (Fig. 7.30). Other

Figures 7.29 Construction of an elevated, sandy breeding area (Turtle Garden) at the former Rand's Marina along Seven Bridges Road.

marinas and boat basins are common along Tuckerton Creek, including several reachable from Seven Bridges Road (Skinners Marina, South Port Marina, Total Marine).

Dredged Channels and Lagoon Development

With various dredging programs, humans have wrought changes in the Sheepshead Meadows. During the 1900s, and especially following World War II, the growing economy and Garden State Parkway construction hastened coastal development, primarily the housing market. Lagoon developments became commonplace, including at nearby Mystic Islands and along Seven Bridges Road. In fact, Mystic Islands, located at the Sheepshead Meadows' northwest border, became among the largest at the time, with approximately 4,500 homes built beginning in the 1950s. These homes, and those in nearby Tuckerton Beach, suffered extensive damage during Superstorm Sandy, but many underwent rebuilding higher and stronger, anticipating another such event with increasing climate change. Over the years, constructed lagoons appear directly along Seven Bridges Road, based on aerial images. Those

Figures 7.30 Aerial view of the Turtle Garden inside the fenced in area at the former Rand's Marina. Note how the surface of the site is being colonized by vegetation. It has already had successful nesting at this new protected location, as indicated by the cages over individual nests.

Figure 7.31 Aerial view of the marshes around the bridge over Big Thorofare (upper center) with lagoon development along the road.

lagoons within the Tuckerton Creek drainage, and adjacent to Holly Lake, were dug and houses constructed between 1931–1956. Those closer to the creek's mouth, at Tuckerton Beach, appear somewhat later, between 1956 and 1963. Another smaller lagoon set appears within the original Willis Creek watershed on the road's opposite side (Fig. 7.31). These lagoons included a connection to Big Thorofare with a long linear ditch stretching between the Thorofare and the lagoons.

Unfortunately, lagoon developments come with numerous ecological impacts. First, they replace productive marsh lands with dredged channels and bury the other marsh portions with the excavated channel sediments. These marshes provide many ecological services, at no cost to the taxpayer, including water purification, buffering against storm-wave damage, wildlife habitat, and biodiversity preservation (Table 7.1). Second, these impacts also eliminate the fish and crab nursery habitats that remain central to these marshes, including those along Seven Bridges Road. Third, detailed studies elsewhere in nearby Barnegat Bay have clearly shown that these dead-end channels, typical of lagoon developments, produce hydrogen sulfide that causes a negative response in many organisms, which helps account for their reduced abundance. Fourth, these developments feature hardened and bulkheaded shorelines, which eliminate the shallow water so central to these productive estuaries. Fifth, the dredged channel feature slower currents, which facilitates accumulating sediment silting, necessitating frequent dredging.

Sheepshead Meadows contains dredged channels for several reasons. One linear channel cut through the meadows along Seven Bridges Road connects the Atlantis lagoon development, once part of Mystic Islands, with Big Thorofare and Little Egg Harbor (Fig. 7.31). In another instance, a larger, deeper channel was dug to eliminate the Little Sheepshead Creek oxbows, providing direct access between the Crab Island fish factory with an associated offloading/weighing facility on Seven Bridges Road (Fig. 4.23). This arrangement allowed fish meal and other products, such as fertilizer made from bunker that the fish factory produced, to be carried by boat to the weigh station where it was offloaded onto trucks, weighed and

Figures 7.32–33 (Left) Helicopter image from May 2017 indicating two confined disposal facilities (CDFs) and a portion of a third (lower right) between Seven Bridges Road (upper right) and a creek and Tuckerton Cove east of Tuckerton. The perimeter berms with brown common reed contrast with the green of the marsh at this time of the year. The uppermost confined disposal facility has a brown interior dominated by common reed. The linear features in the marsh, on both sides of Seven Bridges Road, are mosquito ditches. Note the portion of the lagoon development on Tuckerton Creek (center, left) of the middle CDF. (Right) The confined disposal facilities (CDF) closest to Tuckerton (lower one in Fig. 7.33) is surrounded by a berm dominated by tall common reed in this drone image from March 2019. Note the berm is broken by a ditch into the interior of the CDF. The interior is dominated by marsh grasses similar to those outside the CDF. Seven Bridges Road (upper right) is dominated by red cedars that give it the dark green color. Drone image by Rob Auermuller.

transported along the road to various markets. Historic aerial photographs indicate this channel first appeared sometime between 1931 and 1956.

Dredge Sediment Disposal Placement Sites

Maintaining the channels between Tuckerton Creek and the lagoon developments, including Mystic Islands, requires frequent dredging. The three dredge spoil disposal sites along Seven Bridges Road are located close to the town of Tuckerton (Fig. 7.32–7.33). These sites are easily identified with LIDAR imagery due to their elevation above the marsh surface (Fig. 7.19).

The one closest to Tuckerton was originally constructed on the marsh between 1956–1963 while the other two originated between 1972–1986 based on historical aerial images. The interior of the two sites closest to Tuckerton still contain marsh grass probably because of the connection to higher salinity water from creek channels through the berms surrounding the interior (Fig. 7.32, 7.33). The site furthest from Tuckerton presumably received dredged sediments, resulting in common reed colonization throughout the interior, thus the differences in coloration between the sites. A continuously flooded marsh surface surrounds this site; perhaps the sediment's weight caused the adjacent marsh surface to

Figure 7.34 Image from March 2019 of active dredge spoil disposal site (see crane) on Story Island for deposition of sediments from lagoon developments along Seven Bridges Road. Drone image by Rob Auermuller.

Figure 7.35 An early dredge disposal site, perhaps for the Intracoastal Waterway. The resulting higher elevation has been colonized by a variety of shrubs and trees and is partially surrounded by a brown wrack line at the transition between the marsh and the dredge disposal site.

subside. Little knowledge exists concerning these dredge spoil disposal sites and their ecology, but they are among the highest features on many salt marshes in the Sheepshead Meadows and elsewhere.

Story Island holds another dredged sediment disposal site (Fig. 7.34), and, it is currently being modified to accept more dredge spoils from the nearby lagoon developments, which suffer from reduced access due to sediment build-up. At the other extreme, another, much older dredge disposal site appears evident on the Sheepshead Meadows' eastern end at the mouth of Little Sheepshead Creek (Fig. 7.35). The relatively large tree stand, visible from the bridges at the end of Seven Bridges Road, renders this site easily identifiable.

Figure 8.1 One of the three types of glasswort that occur in the saltmarsh cordgrass zone along Seven Bridges Road. They are typically green during the summer and shades of red in the fall. Photo by Bob Birdsall and Terry O'Leary.

Chapter 8
Flora and Fauna

The previous attempts to define and describe habitats in the Sheepshead Meadows and adjacent islands, as well as their proximity to Little Egg Inlet, provide a baseline for understanding these habitats and the region's flora and fauna. Using a seasonal context provides additional details for these flora and fauna because the temperature's dramatic seasonal variation, among the most variable in the world, heavily influence both. This variation provides the impetus which influences the flora (timing of occurrence and flowering, leaf and fruit production, etc.), and faunal changes (extensive birds, fish, and other animal migration). Using annual, seasonal, tidal, and day-night observation scales offer further faunal life history insights, especially from larval forms to adults, and how their life history varies.

Flora

The flora (terrestrial and aquatic plants including phytoplankton) found around and on the Sheepshead Meadows and adjacent islands represent the New Jersey coast's least-altered portion. Somewhat surprising, new species are still being described, including two new euglenoid algae from marsh pools, a commentary on our taxonomic understanding concerning these small life forms. At the same time, Rutgers University professors are conducting studies to determine if the salt marsh pools produce Harmful Algal Blooms (HABs) that may inoculate adjacent estuarine waters. The studies use Sheepshead Meadows because large portions host unaltered salt marsh, but the meadows also contain altered habitats (parallel grid ditches, Open Marsh Water Management) for comparison. Beyond these study advantages, our emphasis on macroflora and macroalgae includes observing intertidal saltmarsh grasses, underwater grasses such as widgeon grass and eelgrass, various types of macroalgae, and terrestrial vegetation. Saltmarsh grasses, including saltmarsh cordgrass and salt hay, dominate the meadows (Fig. 7.6). The intertidal community, which stretches over the entire peninsula, also contains several species of glassworts (Fig. 8.1) and sea lavender.

The road's elevation and the salinity of the tidal waters largely control the types of vegetation zones found along the road, including immediate roadside vegetation (often comprising grasses and seaside goldenrod) (Fig. 8.2). The next lower vegetation zone, referred to as the tall shrub zone, extends 3–8 feet below the previous zone. These tall shrubs include bayberry (Fig. 8.3), groundsel (Fig. 8.4), winged sumac, and black cherry. This zone includes red cedar (Fig. 8.5), one of the tallest trees along the road, and woody vines (greenbriar, Virginia creeper) with one of the dominant forms being poison ivy (Fig. 8.6). The next zone lower in elevation is the salt hay/marshelder zone. Saltmarsh cordgrass dominates the area's lowest elevation zones, including the tall form, limited to ditch and creek banks, and the short form, commonly found on the marsh platform (Fig. 7.6). The invasive common reed largely dominates a separate zone, occurring

Figures 8.2–8.4 (Top left) Seaside goldenrod, one of the most abundant flowering plants along Seven Bridges Road. Photo by Bob Birdsall and Terry O'Leary. (Top right) Bayberry occurs commonly along Seven Bridges Road. Photo by Bob Birdsall and Terry O'Leary. (Bottom) Groundsell in bloom in fall along Seven Bridges Road.

Figures 8.5–8.6 (Top) Eastern red cedar (right) and roadside shrubs along Seven Bridges Road. (Bottom) Poison ivy, a too typical plant from the tall shrub zone along Seven Bridges Road, is a deciduous vine with clusters of three shiny leaflets that turn red in the fall. Photos by Bob Birdsall and Terry O'Leary.

in dense patches along the road's highest elevations. This species continues to increase in abundance at the Grassle Pocket Marsh (Fig. 8.7). While first described during the 1970s, these separate vegetation zones still generally persist today based on current dominant species observations.

The color of the vegetation along the road distinctly varies seasonally. The marsh greening begins during spring and maintains that color through the summer. At the latter season, the white blooms of Queen Ann's lace conspicuously appear immediately adjacent to the road. During fall, the marsh turns distinctly golden and the marsh grasses flower, most noticeably the tall form of saltmarsh cordgrass. Other fall marsh vegetation blooms include sea lavender, and the glassworts turn a brilliant scarlet red (Fig. 8.1). Some of the shrubs, especially groundsel (Fig. 8.4) feature showy white blossoms. The trees and vines along the road begin displaying their fall colors. The tree leaves eventually drop, while the dominant red cedar remains dark green year-round. Also, during the fall, the goldenrod turns a striking yellow (Fig. 8.2). By winter, the marsh and its common reed displays a dull brown. (For more details, see Chapter 9, "Reading the Road.")

Widgeon grass is the only vascular plant commonly occurring underwater in the Sheepshead Meadows. This species is unevenly distributed among marsh pools, where it can be dominant (Fig. 8.8) or totally absent. An opportunistic species, widgeon grass responds positively to disturbance. It is likely that waterfowl transport the plant, causing its appearance in separate marsh pools, based on studies. Other pool flora can include submerged or floating algae, primarily green fleece, and blue green algae, albeit temporarily variable due to dispersal when high tides and storms inundate the pools. Eelgrass, once abundant throughout New Jersey's Atlantic coast, is no longer evident south of Barnegat Bay. The disappearance from Great Bay, Little Egg Harbor, and around the Sheepshead Meadows may have been due to "wasting disease" in the 1930s and is a good example of a "shifting baseline." However, eelgrass blades are commonly captured along the Sheepshead Meadows' Great Bay side. These blades likely arrive from Little Egg Harbor and Barnegat Bay through the meadows' thorofares. Experiments with artificial eelgrass over the sandy bottoms within the Rutgers University Marine Field Station boat basin clearly demonstrated

Figure 8.7 Common reed growing at higher elevations at the Grassle Pocket Marsh along Seven Bridges Road.

Figure 8.8 Widgeon grass in a marsh pool where it produces oxygen as evident by the bubbles on the leaves. Photo by Thomas Johnson.

that the fish, crabs, and shrimp varieties and numbers far exceeded those observed in the unvegetated areas. These patterns suggest that eelgrass restoration around the Sheepshead Meadows, and elsewhere in southern New Jersey, might enhance the area's nursery value.

Macroalgae comprises a dominant subsurface vegetation form with sea lettuce being the most abundant during the summer and fall around the Sheepshead Meadows and the adjacent islands. Its presence and productivity are evident when it washes up on beaches and tidal flats and when it is abundant in bottom trawls. Other microalgae forms that can be abundant include hollow green weed, Agardh's red weed, graceful red weed, and filamentous sea weeds. Their abundance creates a prominent impact because they can form thick mats on the bottom in nutrient rich areas. Green fleece, an invasive macroalgae originally introduced from Japan via Europe into New York Harbor, has since spread and now occurs in and around the Sheepshead Meadows waters.

Fauna

Observation provides the basis for learning about major faunal (animal) groups. We know more about the visible animals and less about those not easily seen. The first group includes terrestrial species (e.g. birds, and terrestrial and air-breathing aquatic mammals such as seals or dolphins) or those easily visible at low tide (e.g. mussels, worm tubes). Those consistently living beneath the surface, e.g. fish, crabs, shrimp, jellyfish, are more difficult to see, and so detection requires a variety of sampling gears (Chapter 6). This detection difficulty explains, to some degree, why previously undescribed species and new species are still encountered. In addition, some basic taxonomic confusion remains for some groups, but resolution continues. For example, imagine my surprise when we collected a fish species I described as new several years ago from the continental shelf and it recently appeared in our boat basin traps. The inquiline snailfish is a small species (up to 2–3 inches) that typically lives in the sea scallop but leaves them to feed at night. It moves away from scallops and into shallower waters in the spring to spawn, and this, perhaps, accounts for its occurrence in the Great Bay's shallow waters.

Several major habitat types highly influence the fauna common to the study area. The fauna found on the marsh surface and in the pools, both on the Sheepshead Meadows and the adjacent islands, is fairly resident and lacks diversity, but remains important within the ecosystem because of their abundance. The diversity increases dramatically in the creeks and thorofares, with a distinctly seasonal component to match the differences in water temperature, some of the world's most extreme. Seasonal migrations and life history related movements reflect these findings, especially for the fish and some crabs. The fish passing under the road's bridges are diverse based on tides and seasons. The terrestrial fauna found along the road can be diverse, especially due to migratory birds and some insects moving seasonally along the East Coast. Little Egg Inlet's proximity greatly influences the varied fauna inhabiting the marshes, adjacent thorofares, and bays. Extensive daily and seasonal migrations also occur bringing a faunal array, including fish in all life-history stages, horseshoe crabs, bottlenose dolphins, and harbor seals.

Easily Observed Fauna

Marine and Terrestrial Mammals

These distinctive mammal groups comprise common faunal components on the Sheepshead Meadows and adjacent waters. Terrestrial mammals are often seen along Seven Bridges Road or dead on the road from nocturnal vehicular traffic during periods of heightened animal activity. Traffic victims include opossum, mink, red fox, raccoon (Fig. 8.9), and an occasional white-tailed deer (usually on the road's Tuckerton end). The numerous raccoons are more common than we prefer because they occupy our buildings, including in the walls, and cause innumerable problems, especially because they have learned how to avoid set traps. River otters occasionally haul out onto our boat basin's floating docks when passing through the neighborhood. The dam and the fish ladder that leads from Tuckerton Creek into Lake Hopatcong provides another place to observe otters. Mink, both live and dead, appear on Seven Bridges Road and use our boat docks and boats as shelter, making them easily observable. In fact, we have observed mink every year on the road or at the Rutgers University Marine Field Station since 2013. Red fox appears fairly consistently from the Station and along Seven Bridges Road since 2014. Other mammals, including gray squirrels and groundhogs sighted on the road, are likely drop-offs that people trapped in their yards and then released on Seven Bridges Road. A well-documented report places a black bear on the nearby Atlantis Golf Course, sighted on the uplands adjacent to the marshes near Big Thorofare.

The marine mammals around the Sheepshead Meadows are typically spotted during the day. These include the seasonally abundant bottlenose dolphins (Fig. 8.10), which are most evident with the warmer summer water temperatures. These migrate

Figures 8.9 Mammals found along or seen from Seven Bridges Road including red fox (photo by Birdsall and O'Leary), raccoon (photo by RUMFS staff), road kill mink (photo by author), and river otter (photo by RUMFS staff).

into the area outside Little Egg Inlet and into the deeper waters behind Little Egg Inlet from their overwintering grounds below Cape Hatteras, North Carolina. A good example of this is an individual dolphin, called #56, freeze-branded decades ago in Florida when few options existed for tracking them. Several of our crews frequently observed this dolphin around the inlet, but it then disappeared. Another marine mammal that has wandered into our area from Florida waters is the manatee, observed in Great Bay several years ago. There was concern about its demise, but it was found along Florida's east coast in 2022, when it received a new satellite tracking tag. This animal is believed to be about 35 years old and could occur here again, so keep watching for it.

Harbor seals are another seasonally abundant marine mammal commonly viewed from the Rutgers University Marine Field Station and in Shooting Thorofare, where they rest on sand bars at low tide (Fig. 8.10).

Figure 8.10 Marine mammals common around the Sheepshead Meadows including bottlenose dolphins (top) and harbor seal (bottom) resting on nearby Fish Island. Photographs by Jackie Sullivan.

Although they can be observed resting in various Great Bay locations, we find them most easily when they haul out on Fish Island, one of the Seven Islands group. We keep an observation log, and typically see them from November through about April during most years, before they migrate back north. Over this period, they remain abundant, based on our spotting-scope observations from the Rutgers University Marine Field Station porch. In recent years, the numbers continue to grow; our count totals 265, while a Stockton University camera system identified 330 at the same time on March 7, 2022.

Birds and Their Migrations

The bird diversity observable from Seven Bridges Road in the Sheepshead Meadows include residents and migrants, both taking advantage of the natural and man-made habitats. The Sheepshead Meadows location along the Atlantic Coast Flyway accounts for the consistent and detailed seasonal bird migrations. The red knot and greater yellowlegs comprise the best known for their extensive seasonal migrations (Fig. 8.11). Prior selected species studies have documented the seasonal occurrence quite well during migration. These species use the Sheepshead Meadows' marshy tip adjacent to Shooting Thorofare at the end of Seven Bridges Road (oystercatcher, willet, ruddy turnstone, black bellied plover, greater yellowlegs, least sandpiper, dowitcher, whimbrel, and sanderling). Birders have photographed other species, especially wading birds (great egret, snowy egret, great blue heron, little blue heron, and tricolored heron) elsewhere along the road (Fig. 8.12), a frequently visited birding hotspot.

Figures 8.11 Both the red knot (left) and greater yellowlegs (right) make extensive migrations from southern South America to the Arctic with New Jersey, including Seven Bridges Road, as a stopover point. Note the habit of the latter to stand on one leg. Photo by Robert Harvey.

Fish-eating birds (osprey, bald eagle, and belted kingfisher) comprise another group in this productive area (Fig. 8.13). Our observations along the same road indicate these birds are seasonal migrants, including osprey, harriers, great and snowy egrets, and glossy ibis. Tree swallows appear in large numbers annually during autumn (Fig. 8.14). Barn swallows, during their seasonal migrations, are especially easily observed, since a large colony nests under the Rutgers University Marine Field Station buildings. Some more obvious winter species along the road include brant as they flock in Shooting Thorofare or in large pools on the marsh surface at the end of Seven Bridges Road. The same can be said about the frequently numerous black ducks.

Recent individual bird tracking studies verify the broad geographical migration extent for some species. Researchers using coordinated radio telemetry established the MOTUS Wildlife Tracking System as an international, collaborative network. The Rutgers University Marine Field Station hosted a deployed MOTUS receiver from 2017 through 2021 (Fig. 6.3). The MOTUS program, while at the field station, detected species from the Canadian Arctic to Brazil from ten locations (Fig. 8.15). The tagging locations vary with species and seasons, but they include sites as far north as Churchill, Manitoba, Canada (1600 miles from the Rutgers University Marine Field Station) and as far south as Caicara, Brazil (4,700 miles from the field station). The twelve red knots detected at the field station were tagged primarily at Delaware Bay in the fall as they migrated south for the winter, but some also received their tags in nearby Brigantine, New Jersey. These were also detected in the fall, typically in the same year and month as tagged. A few others obtained their tags in South Carolina during the spring and were detected two weeks later at the Rutgers University Marine Field Station, presumably during the spring northward migration. For the semipalmated plover, the seven tagged birds all received tags during June of several years in Arctic Canada, most in Churchill, Manitoba, or on one occasion in James Bay (in August) (Fig. 8.15) with all detected in August at the Rutgers University Marine Field Station on their fall southward migration.

For the semipalmated sandpiper, fourteen birds received tags at various locations in the spring, ranging from James Bay, Miscou Island, New Brunswick, the St. Lawrence Seaway and elsewhere in Atlantic Canada (Fig. 8.15), but detected during fall the same year at the Rutgers University Marine Field Station. The most distant tagging sites tracked at the field

Figures 8.12 Array of typical bird species found along Seven Bridges Road as migrants: semipalmated sandpiper (upper left), oystercatchcher (upper right), snowy egret (middle left), great egret (middle right), black crowned night heron (lower left), and a year round resident great blue heron (lower right). Photos by Robert Harvey.

Figures 8.13 Typical fish eating birds photographed along Seven Bridges Road including osprey with Atlantic needlefish (upper left), belted kingfisher (upper right), young great blue herons in nest (middle left), little blue heron (middle right), tricolored herons (lower left), and common loon (lower right). Photos by Bob Birdsall and Timothy O'Leary.

Figure 8.14 Swarm of migrating tree swallows in the air above First Bridge over Big Thorofare that typically occur in the early fall. Photo by Bob Birdsall and Timothy O'Leary.

station were two semipalmated sandpipers from Caicara, Brazil, tagged in February and March and subsequently detected together in late May at the station. Birds tagged in the spring in the United States either originated from Delaware Bay or Jamaica Bay, New York. The Rutgers University Marine Field Station detected these birds days later, for Delaware Bay tagged birds, or during the fall for most of those coming from Jamaica Bay. One tree swallow was tagged in Amherst, Massachusetts, in late June and detected at the Rutgers University Marine Field Station in late August, perhaps among the flock flying over Seven Bridges Road (Fig. 8.14). Another tree swallow was tagged in New Brunswick, Canada, near Prince Edward Island on July 3, 2019, and then detected 31 times from July 21 through August 26, 2019, at the Rutgers University Marine Field Station.

For many other birds, human-created habitats can provide easy observation locations for spotting typical birds along the road. The roadside patches of common reed, which grow there due to the road's original construction, often serve as springtime sites for redwing blackbird displays, and the road itself serves as a feeding area when insects are abundant there. The power lines, both active and those abandoned with only poles remaining,

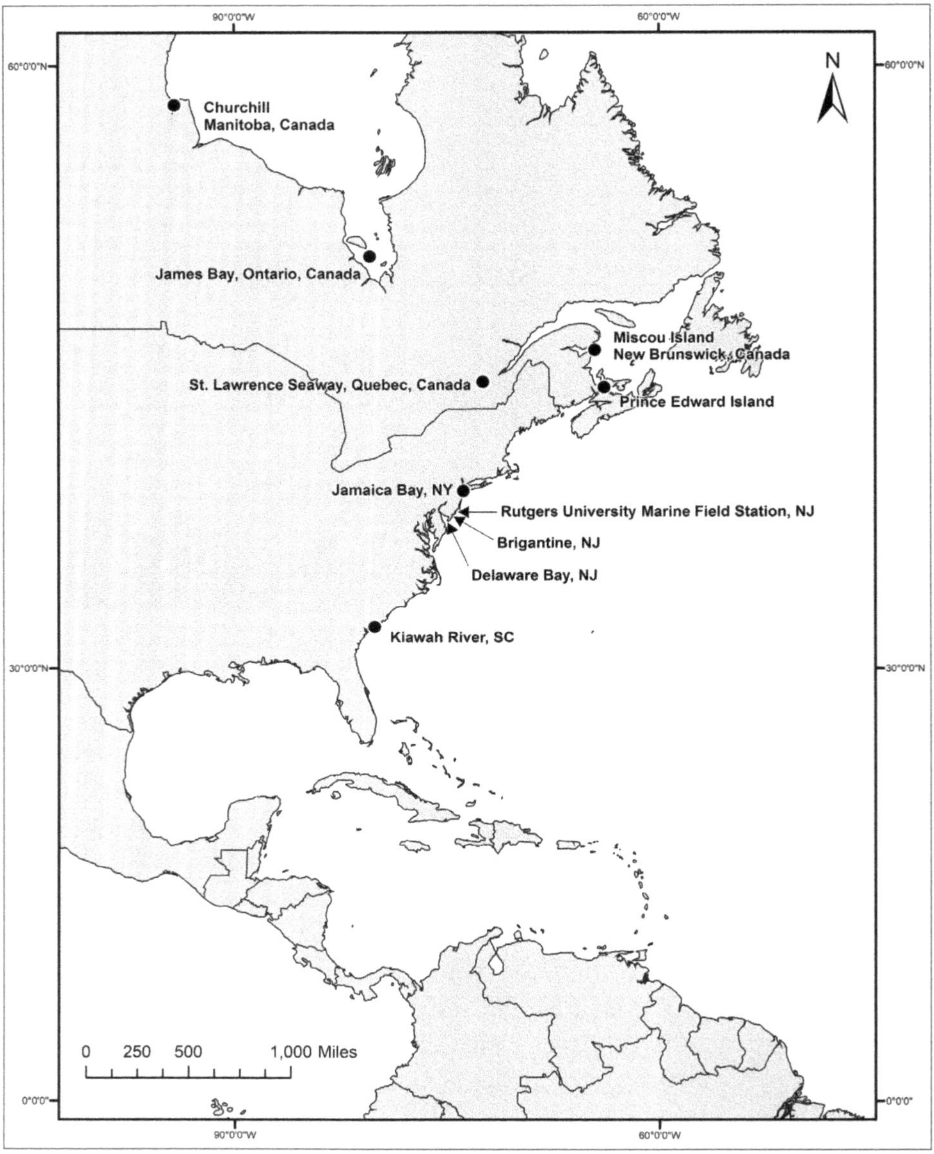

Figure 8.15 Bird detections locations throughout North America based on individual tagged birds that were also detected at the Rutgers University Marine Field Station including red knot, semipalmated plover and tree swallow. Another site not represented on the map is Caicara, Brazil.

Figures 8.16 Large osprey nest on power line pole in Newmans Thorofare near Crab Island Fish Factory (left) and eagles nest on power line in marsh crossing the eastern end of the Sheepshead Meadows (right). In the latter we are examining eagle prey items found under the nest. Photos by author and RUMFS staff.

provide perches for some birds. These include belted kingfishers on the power lines, especially those over or near marsh pools or over thorofares, for feeding. The lines also serve migrating tree swallows as resting spots and a variety of black birds during the fall each year. Individual poles serve as feeding perches for osprey as they pick apart menhaden captured from local waters. Some of these poles also provide osprey nesting sites during the summer and occasionally bald eagles in the winter and spring (Fig. 8.16). The Turtle Garden at the former Rand's Marina is a common location for predators seeking the abundant, recently hatched diamondback terrapins there. Another large bird that can be observed in the Sheepshead Meadows is the mute swan, an invasive species from Europe (Fig. 8.17). Observed in Tuckers Creek at the Grassle Marsh, this species and its young seem to prefer the larger open water bodies of water covering the marsh surface, such as the "lake" we refer to as "Swan Lake" along Seven Bridges Road where they frequently feed.

Figure 8.17 Invasive mute swan in Grassle Pocket Marsh with young on back and in the water.

Turtles

The sole turtle that is resident in the salty waters of estuaries such as those surrounding the Sheepshead Meadows is the diamondback terrapin. It completes its life cycle here, observable from a boat as they pop their heads above the water to breathe. Seen primarily in the spring through fall, they hibernate during winter in the creek mud. When the females move to sandy higher ground to lay their eggs, vehicles traveling the road often strike and kill them, making their carcasses easily observable. The female turtles use the tides to time their arrival at the nesting sites, with the greatest number observed on and around the new and full moons which produce the highest tides. At this time, they can swim rather than walk through the marsh vegetation to reach the road. On occasion, the female numbers on or crossing the road can be extraordinary, with 53 on one drive and 87 on another between Tuckerton and Shooting Thorofare.

Several detailed studies have determined the mortality rates and locations. As a result, we now know vehicular traffic plays an important role in these mortalities on Seven Bridges Road with the roadway's developed portion to the First Bridge logging the highest deaths. Road mortality varies during the 24-hour cycle, with morning revealing the greatest number, indicating most mortality occurred overnight. This may result from frequent nocturnal nesting activities, but it also occurs through the day. One nesting location that has reduced road mortality is the sandy beaches at the end of Seven Bridges Road. This has been especially obvious at a point near the Rutgers University Marine Field Station where a washover beach has been forming for some time (Fig. 7.2, 7.3). At this site, numerous females swim off the beach in the boat basin, with some making their way onto the sand, exploring several sites, and eventually constructing a nest, laying eggs, and burying it before moving back into the water. After hatching from the eggs deposited along the road, the hatchlings scamper into the nearest vegetation all along the road and disappear for several years. Some scientists refer to this time as the "lost years," apparently comprising the time they spend on the marsh surface, often under wrack, where they feed and grow, based on studies elsewhere. Extensive all-season sampling, tidal stages, and multiple water habitats with multiple gears over decades provides the evidence supporting this interpretation for the Sheepshead Meadows (Chapter 6). Numerous publications and two books of mine summarize this activity with not a single hatchling observed in the water, suggesting strongly that their nursery habitats lie on the marsh surface. Furthermore, laboratory observations using terrariums with muddy substrates frequently found them buried in the mud.

Notes from the Road—Bayley's Long Journey Home: A Terrapin Poacher's Dilemma

A female diamondback terrapin, named Bayley through a contest for local school children, experienced a heart-wrenching story based on her travels by illegal poachers. She originally received her tag and was released during a June 2008 study of the natural populations along Seven Bridges Road administered by researchers from Drexel University. At that time, she was seven years old. Beginning in 2016, Governor Chris Christie declared it illegal to harvest terrapins, the only turtle that completes its entire life history in New Jersey estuaries.

Special agents from the US Fish and Wildlife Service Office of Law Enforcement, in collaboration with the New Jersey Division of Fish and Wildlife Enforcement and the Endangered and Non-Game Species Program, unfortunately found that an illegal poacher had taken Bayley, and many other terrapins, along Seven Bridges Road, based on an extensive examination in 2017. Once apprehended, the poacher was prosecuted and sentenced to six months in prison, six months' probation, and a $250,000 fine for possessing over 3400 diamondback terrapins, both adult females and hatchlings, at his Pennsylvania home.

Little information exists about Bayley's time in captivity, except that she was a victim of the illegal pet trade and possibly used in captive breeding. Her tag, however, provided some details, including that she spent time in Tennessee, Massachusetts, and eventually Maine, where her tag finally identified Bayley, and the authorities transported her back to New Jersey, where The Nature Center at Island Beach State Park sheltered her. In 2013, Ben Wurst, of the Conserve Wildlife Foundation of New Jersey, and others decided she should be quarantined and evaluated for her health prior to her release back into the wild. Once proven healthy, she was ceremoniously released back along Seven Bridges Road to continue her life in the natural habitats there. Since then, and thanks to many individuals, she is safer because many people engage in conservation and educational activities along Seven Bridges Road. The most recent safety measures are roadside signs that notify drivers about potential road mortalities due to numerous females laying eggs along the road.

Bayley when released at the former Rand's Marina near the Turtle Garden on Seven Bridges Road.

The only other turtle to appear in the Sheepshead Meadows vicinity is the loggerhead. One large adult, captured in our trawl sampling, was notable because it had a piece of plastic wrapped around one of the front flippers. After removing the plastic we notified the Marine Mammal Stranding Center in Brigantine which then released it. Another smaller juvenile washed up dead and decaying on a beach near the Rutgers University Marine Field Station. Their occurrence just inside Little Egg Inlet is no doubt the result of seasonal oceanic migrations up and down the United States' East Coast.

Horseshoe Crabs

Not true crabs, horseshoe crabs are more closely related to spiders. These ancient animals occupy the shallow, sandy shorelines around the Sheepshead Meadows and the adjacent islands. The adults only occur seasonally, during late June through early August, when they migrate inshore from the ocean to spawn, typically on sandy beaches closest to the inlet (Fig. 8.18). They come ashore, preferentially on high tides, to lay their greenish eggs

Figure 8.18 Adult horseshoe crabs accumulate on spawning beaches during the early summer to deposit their eggs. Typically, the male attaches to the tail of the female in order to fertilize the eggs.

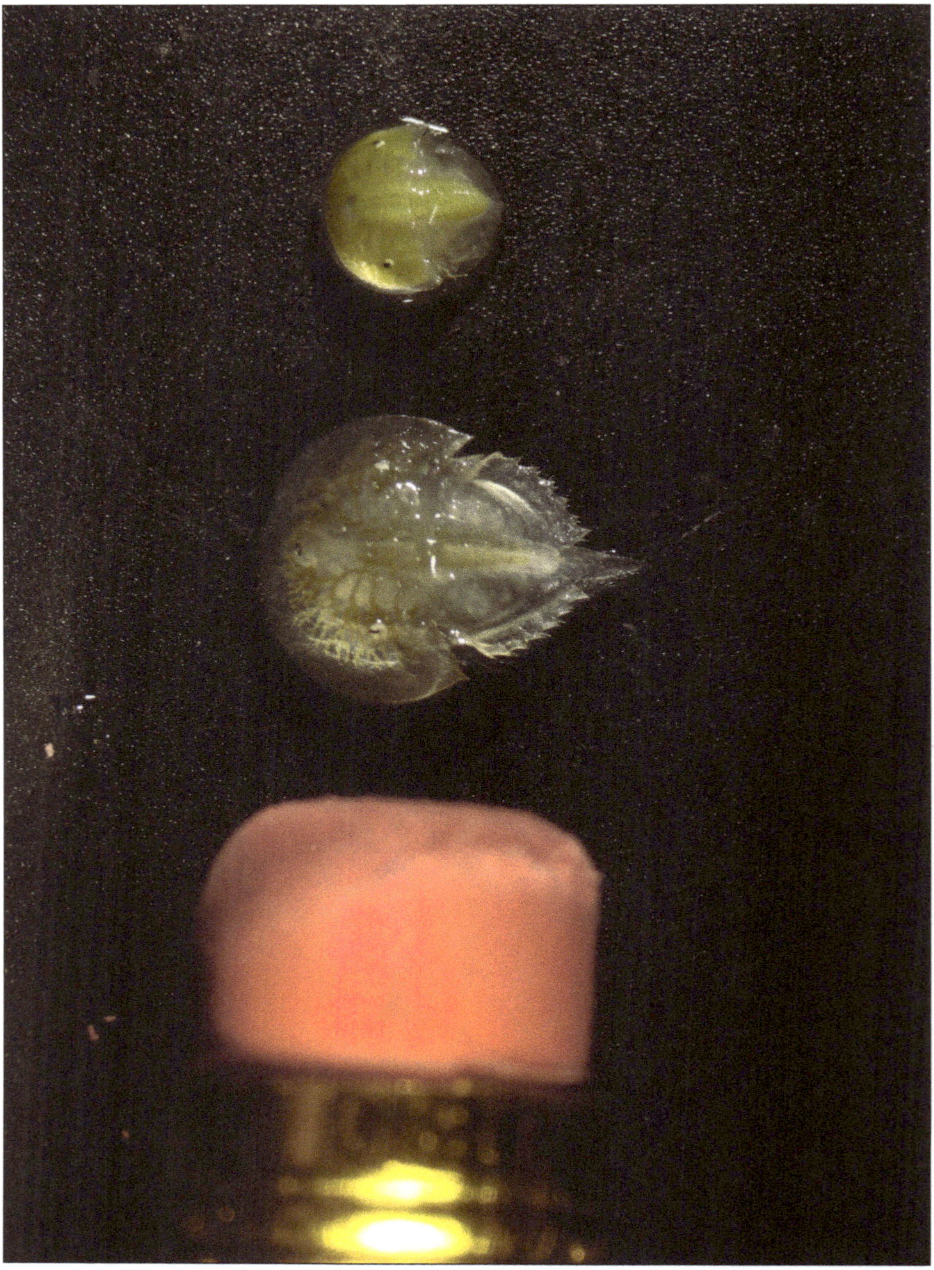

Figure 8.19 Early life history stages of horseshoe crabs are regularly collected in the water column of thorofares stretching through the Sheepshead Meadows. The first and smallest stage is the trilobite stage. Subsequent stages resemble miniature adults (with pencil eraser). Photo by Paola Lopez-Duarte.

in the sand. Some adults get stranded and die, and their carcasses litter the beaches. During early development stages (typically less than 1/8 inch and without a tail), they undergo changes in morphology as they grow until they resemble miniature adults (Fig. 8.19). The larvae live in the water column where we capture them in our regular night-time plankton sampling from June through August, with the peak in July, until they move to the bottom. They can be found there, although not easily, because they burrow into intertidal and subtidal sandy habitats for several years. When they reach about four inches in width they migrate offshore, outside Little Egg Inlet into the ocean, and feed and grow there until they are adults. They then migrate back inshore to reproduce. They repeat this pattern over 17–19 years, based on our own and other studies.

Harder to Observe Fauna

Some faunal components in the Sheepshead Meadows area are either small, remain hidden among the vegetation (insects and spiders), or they live underwater, making them all difficult to observe. Such fauna must be captured in order to even realize their presence (fish, shrimp, crabs—particularly the small, early life history stages).

Insects and Spiders

The Sheepshead Meadows insect and spider populations are the most frequently and thoroughly studied faunal components, thanks to the late Bob Denno, who spent nearly 40 years extensively examining these creatures. He began his career in the Barnegat Bay's Manahawkin salt marshes as a Rutgers University postdoctoral fellow. At that time, he likely recognized the significance found in the natural marsh habitats along Seven Bridges Road and that prompted his life-long research interests there while at the University of Maryland. Numerous graduate students, postdoctoral fellows, and colleagues from the University of Maryland and elsewhere have supported the breadth of his studies. The "Denno School" has produced outstanding publications based on a sound understanding of natural history and salt marsh insects (especially planthoppers), spiders, and other arthropods along Seven Bridges Road. These "ecology of place" studies at this location included plant-insect interactions, dispersal, interspecific competition, predator-prey interactions, and food web dynamics. His former students and colleagues continue these basic ecological studies as I write this today. These scientists currently maintain an emphasis on the impacts of habitat fragmentation on marsh plant and insect population genetics, nutrient impact on salt marsh community, and ecosystem dynamics, as well as beneficial use of dredge spoils on the same communities. While many small insects are hard to notice while traveling along the road, the webs of the abundant spiders, whether orb weaving, sheet web building, or hunting spiders, can easily be seen, especially on a foggy, damp day (Fig. 8.20).

While many insects associated with marshes are harder to observe, some want to find us. Among these, the most prominent are greenheads and mosquitoes. Some biting greenheads were poorly known taxonomically and included one undescribed species. As a result, we have only recently begun to understand their life history. We do know that adult greenheads mate in the air above the marsh. The female subsequently lays her eggs in the mud, and the larvae develop there where they feed on invertebrates. They overwinter in the same habitat, develop into a pupa in the spring, and hatch into the adults in summer. As adults, the males feed on plant nectar but the females require a mammal's blood meal to provide for broods after the first one. This blood meal can be from any mammal, including us. This is most obvious when people walking, jogging, birding, etc., along Seven Bridges Road become distracted by their biting attacks. Regulars are usually prepared, but the first-time tourists to the road and surrounding

Figure 8.20 Two types of spider webs (previous page, orb-weaving) and (above, sheet web building) that capture prey that feeds on marsh grasses and algae/detritus along Seven Bridges Road.

marshes often are not. When novices pull up at bridge stoplights, if they leave their windows open, greenheads quickly attack, in part because the carbon dioxide exhaled by the occupants and because of the vehicle's heat. This leads to frantic swatting at the invaders inside the vehicle until the light changes, if they notice, and they try to chase the greenheads from their vehicle as they proceed down the road.

The mosquito fauna along Seven Bridges Road is changing. The eastern saltmarsh mosquito is the most common and biggest nuisance species. It typically lays its eggs on mud in the saltmarsh and the eggs delay hatching until tidal inflow or rainfall floods the eggs, where they develop in the water. Once the larvae (wrigglers) hatch, their development lasts about 5–7 days until they change into the flying, biting adults. A new, closely related species, the black salt marsh mosquito, from more southern parts of the United States, has been extending its range into New Jersey, including along Seven Bridges Road with warming temperatures. It has a similar life history to the eastern saltmarsh mosquito except that the wriggler stage may last a few days longer. It is not, fortunately, as abundant as the previously described species. As a result of these species' nuisance factors, the Ocean County Mosquito Commission has made numerous attempts to control these blood-sucking pests in the Sheepshead Meadows area. These attempts range from salt marsh ditching in the early 1900s to Open Marsh Management into the 1990s. Today, these efforts continue by monitoring the larvae and determining wriggler abundance by dip-netting known mosquito breeding areas (see Fig. 1.7 for locations).

Nocturnal moths are some of the least frequently observed insects along Seven Bridges Road. In an ongoing study, scientists use a white, lighted sheet to attract the moths and photograph them from the deck of the Rutgers University Marine Field Station. During two collecting events in the summer of 2020, nearly 50 moth species appeared. One of the most striking was the pandorus sphinx (Fig. 8.21), a species that occurs throughout the eastern United States, but little is known of its specific habitat use.

The monarch butterfly is quite conspicuous in the Sheepshead Meadows appearing during the fall migration. They are most abundant in late August or September through November. At this time, they occupy the tall vegetation at road's end near Shooting Thorofare (Fig. 8.22), but they can appear all along the road either flying through or resting in the vegetation. This occurrence is part of their spectacular movements along the coast on their way to their Mexican wintering grounds, where they gather in communal roosts until spring when they migrate back north again.

Figure 8.21 Pandorus sphinx moth photographed under lights on the deck at the Rutgers University Marine Field Station by Ann-Marie Woods and Tom Grothues.

Figure 8.22 Monarch butterflies resting in vegetation at the end of Seven Bridges Road during the fall migration to the south for the winter.

Shellfish

A variety of shellfish have been historically important and continue to be so in the Sheepshead Meadows area today. The decline of oysters along the coast of New Jersey is well documented. In the heyday of oysters and the fishery for them during the late 1800s, they were abundant, and people assumed the fishery would last forever. But a combination of poor management, disease, and habitat destruction along the coast, due to urbanization, contributed to their decline such that the fishery was limited by the 1950s. The diseases MSX and Dermo in the 1950s and 1960s drove another nail in the coffin, further reducing the natural oyster set, including around the Sheepshead Meadows and elsewhere in the state and all along the East Coast. An effort to recreate/restore/evaluate an oyster reef, the Tuckerton Reef, in Little Egg Harbor is one effort occurring near the Sheepshead Meadows. Multiple individuals and institutions between 2016 and 2019 began using oyster larvae from a disease-resistant line set on whelk shell in large outdoor settling tanks of Great Bay Marina near Graveling Point. These sets, and wild set transplants from the lower Mullica River, were deployed at the Tuckerton Reef and evaluated over three years. The Tuckerton Reef's location may be especially appropriate because Barnegat Bay modeling studies have demonstrated a general northward transport of oyster and clam larvae from waters near Little Egg Inlet to the bay's northern part.

In the last decade numerous indications show that the oysters are recovering. One index is the number of oysters we see that have set and are growing on intertidal ribbed mussels on Crab Island marsh edges (Fig. 8.23), and also at Willit Creek near the Station. In addition, juvenile oysters (spat) have set on our traps in the boat basin (Fig. 8.24). Larger oysters are evident in the Station boat basin, both on the boat railway and the dock pilings. The reason for these apparent increases is not completely understood. One likely explanation is the increasing number of oyster culture operations in nearby Great Bay and adjacent Little Egg Harbor, which allow more reproduction and thus more larvae in the water column that can set on available hard substrates. Also, the increasing use of living shorelines has increased the amount of hard substrate available on which oyster larvae can set. Regardless of the responsible factors, an important

Figures 8.23–8.24 (Left) Oysters growing on intertidal ribbed mussels on edge of salt marsh on Crab Island. (Right) Oysters that have set as larvae (spat) and are growing on fish traps in the Rutgers University Marine Field Station boat basin. Photo by Roland Hagan.

faunal component of the estuarine ecosystem appears to be returning. This species and its fishery are a good example of shifting baselines over the last 150 years. Some of this is relived in the documentary *The Oyster Farmers* and related background material in the Summer 2017 issue of *Edible Jersey*, which focuses on oysters and the growing oyster aquaculture industry.

The only time you can easily see subtidal hard clams that bury in the bottom is when they are preyed upon by herring gulls, most often in the winter, unless of course you collect them yourself. The surface of Seven Bridges Road provides a long, hard surface that these gulls take advantage of when they drop the clams high over the road and eat them after the shells are broken open. However, our deck over the water is often littered with their broken shells. Another common shellfish in the area is the intertidal ribbed mussel. These are abundant along all of the marsh edges particularly in the creeks (Fig. 8.23). They are also preyed upon by herring gulls and their shells are abundant on Seven Bridges Road at the same time that hard clams are found there.

Crabs and Shrimp

Several native crab species can be observed in and around the Sheepshead Meadows (Fig. 8.25). The blue crab, a swimming crab, receives the most attention due to their importance in recreational and commercial fisheries. They have a complex life history with the females fertilized while in the estuary, then migrating offshore to release the larvae (zoea, at sizes up to 1/10 of an inch) in the ocean, which make their way back into estuarine waters. Once there, they leave the water column and settle to the bottom. When the small juveniles measure up to 1 and 1/4 inches, they can leave the bottom and move to an alternative location. That is when we catch these juvenile crabs in our nets off the bridge in Little Sheepshead Creek, where we take weekly samples on night flood tides to catch larvae of many species since 1989. They occur primarily in the fall. The habitat use and growth of these and larger juveniles is critical to their survival in these estuarine "nurseries." They are especially abundant in the creeks and meadow thorofares, but can also be abundant in the surrounding bays, where they are easily observed in skinny water or as adults in your crab pot. Our studies demonstrate these crabs are most abundant in macroalgae, such as sea lettuce, because they survive well there, based on predation studies. Based on analysis of two 25-year data sets for small (about one inch) and larger (up to 2 ½ inches) crabs taken with our plankton nets in Little Sheepshead Creek

Figure 8.25 Native crabs in the Sheepshead Meadows including (clockwise from top left) blue crab, lady crab, marsh crab, and male fiddler crab.

and traps in the Rutgers University Marine Field Station boat basin, we now have several insights. In general, a positive relationship exists between the small juveniles' abundance and nor'easters, indicating that the storms aid in transport of the larvae through Little Egg Inlet. A positive relationship also exists between the small juveniles' abundance and warmer winters, indicating that moderate temperatures increase their survival.

Another swimming crab, the lady crab (Fig. 8.25), is a lower estuary dweller, closer to the inlet and in the ocean, both featuring sandy beaches and bottoms. Sandy habitats provide the crabs with a burying medium, thus avoiding predators. Despite that ability, they also use the water column from early May to late November as small juveniles, less than one inch, to disperse between habitats, based on our frequent collections under night-lights in our boat basin and occasionally in our plankton nets.

Fiddler crabs (Fig. 8.25) are the third most common and important species of crabs that occupy the salt marsh edges along creeks and thorofares. Here, the three species act as bioengineers, influencing exchange of sediment, nutrients, etc., largely due to their burrowing. Fiddler crabs are easily observed at low tide on the marsh surface as they forage or as the males wave their single large claw around while courting during spring through summer in the Sheepshead Meadows. They are less easily observed in the early life history stages (zoea) after the females release them, and they develop in the lower estuary or in the adjacent ocean and are transported (as megalopa) back into the estuary, where they settle and transform into small juveniles resembling the adults. We have collected the smallest mud fiddlers (one-quarter-inch shell width) as early as May and through August and September 2019 and 2020 in our "pit traps" (glass casserole dishes sunk into the marsh until they are level with the surrounding sediments). Once they moved onto the marsh surface on high tides, we captured them in a small marsh adjacent to the Station boat basin and Willit Creek. We collected over 10,000 juveniles, at sizes up to 15–20 mm. shell width, in these reference and experimental marshes, evaluating thin-layer dredge spoil sediment deposition sites. While our sampling collected fiddler crabs in abundance, we know the number to be greater based on observing Laughing and herring gulls feeding in the pit traps. While the mud fiddler species dominated marsh collections, the sand fiddler appeared in very sandy habitats less than 50 yards away. Other studies demonstrate that these crabs choose different habitats based on sediment type and attraction to areas with adults of their species.

Marsh crabs are a co-occurring species that burrow in the sediments along the vegetated marsh edge in the same marshes (Fig. 8.25). They are nocturnal and seldom seen, even though they are larger than fiddler crabs. They modify the marsh surface because they feed on marsh grass leaves and roots and their burrow openings are larger (Fig. 7.4) than those the fiddler crabs create. As a result, where they are abundant, the marsh creek edges are unvegetated and reveal heavily bio-eroded peat substrates. Besides feeding on the vegetation, some studies suggest they feed on fiddler crabs as well.

Several common crabs and shrimp have invaded the East Coast, including the Sheepshead Meadows area. The green crab, an invasive species from northern Europe (Fig. 8.26), is a common component of the fauna collected in a long-term time-series based on twice a week wire mesh trapping in our boat basin. Between 1990 and 2014, this species frequently ranked within the abundant top ten. During this time, the crab commonly appeared between late fall and early spring, but with a declining population over time. This decline might result from native blue crab predation, and/or increasing water temperatures on this northern species. Another invasive crab, the Japanese shore crab (Fig. 8.26), has frequently appeared in the same traps as juveniles since 1995, where they rank among the top ten species during some years. A theory suggests this species arrived on the East Coast after being transported in oceangoing ships' ballast water. Observations typically in the summer and fall also place the adults and juveniles under hard substrates, pieces of concrete and wood, lying under the Rutgers University Marine Field Station buildings.

There are numerous native shrimp species that occur in the Sheepshead Meadows area based on our sampling programs. They play a critical role in estuarine food webs because they feed on plant material (detritus) and serve as food for many economically and ecologically important fish and crab species. We have identified six native species that differ in the area's temporal use (Appendix Table 1). We determined the most important species based on an intensive one-year weekly sampling with plankton nets at our long-term Little Sheepshead Creek sampling site and trap sampling in the Rutgers University Marine Field Station boat basin.

The sand shrimp was the most abundant species in the plankton samples. These had a distinctly seasonal component with most of the adults (greater than six tenths of an inch body length) collected in January and February, and the smaller

Figure 8.26 Invasive crabs (Top) Green crab, in this case a molting individual (note barnacle on old shell and lighter green of new shell). (Middle) Japanese shore crab, and (Bottom) oriental shrimp. Photos by author except last one by Jim Carlton.

juveniles common from February through August with peaks in May through July. This seasonality apparently reflects their movement into the nearby ocean during the fall and early winter and their return again in spring and summer.

The other abundant shrimp, marsh grass shrimp, dominate the trap samples in the boat basin and also the Little Sheepshead Creek plankton samples. The adults, including those carrying eggs, appeared most abundantly from April through July, but the samples collected during July provided the only juveniles (less than three tenths of an inch). The other much less abundant shrimp species include daggerblade and brackish grass shrimp, and the false zostera shrimp, as well as the burrowing ghost shrimp. Other shrimp species collected in more recent years in our otter trawls throughout the estuary included brown shrimp and white shrimp. Both of these commercially important species have expanded their ranges from more southern areas, apparently responding to regional warming water temperatures. An invasive species, the oriental shrimp (Fig. 8.26), has appeared in our trap and trawl samples since 2017. This species, originating in the western Pacific, presumably arrived in the ballast water of large ships entering ports along the East Coast and spreading from there.

Another crustacean that has surprisingly appeared in the Sheepshead Meadows area is the American lobster, but only in the stage immediately before settlement to the bottom (Stage IV). These larvae, which typically

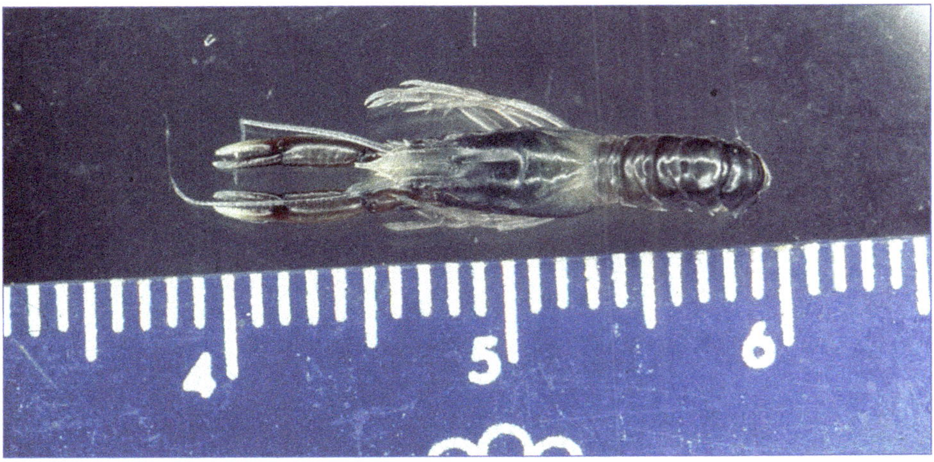

Figure 8.27 The planktonic stage (Stage IV) of American lobsters, before settlement, resemble miniature lobsters when they have been collected under night lights in the Rutgers University Marine Field Station boat basin.

measure one-half inch, resemble miniature adults (Fig. 8.27) when captured under lights shining on the water. They appear most abundant in our nightlight samples during June but occur as late as early July. The larger juveniles have never been collected in our extensive multiple habitat types sampling throughout the estuary, conducted with multiple sampling gears.

Fish

The fish, in all life history stages from eggs, larvae, juvenile to adults, are a particular focus for the scientists, students, technicians, and their collaborators working at the Rutgers University Marine Field Station. Thus, the following extensive discussion builds upon and fills the gaps in our earlier studies, especially relative to the dominant habitats.

Inlet Users

Little Egg Inlet is one of New Jersey's most natural inlets and therefore prominent in understanding the state's coastline. The inlet's waters connect the bays and thorofares to the remaining estuary day and night at all tidal stages, even during storms. Thus, the inlet's waters are central to the aquatic fauna. These may vary on longer time spans, as described earlier, because the inlets open, close, and move on the scale of decades to centuries (Fig. 2.3) and sometimes over just a few days during major storms. We have monitored the various faunal types passing through the inlet. One approach sampled fish and crabs in the shallow, sandy beaches at the former reincarnation of Tuckers Island just inside the inlet and other islands (Tow, Little Beach, Holgate, Goosebar Sedge) in the inlet's vicinity from June 1994 through October 1995. During our sampling, both day and night, sometimes while camping there overnight, we determined the species composition and abundance with a beam trawl and a seine. Of the 59 species collected, recently spawned fish either from the ocean (American sand lance, windowpane, northern kingfish) or in the estuary (Atlantic silversides, winter flounder, striped killifish, northern pipefish, bay anchovy) dominated the catches.

Bays and Channels

We know that other estuarine species migrate through the inlet during their spawning and seasonal migrations, including horseshoe crabs and others that enter the estuaries for feeding, including bottlenose dolphins and harbor seals. These species use the deeper waters between the inlet,

Great Bay, and the Sheepshead Meadows' southern end, including Shooting Thorofare, Grassy Channel, Newmans Thorofare, Marshelder Channel, and the Intracoastal Waterway (Fig. 1.1). Fish commonly collected in the open bay waters with trawls include bay anchovy, Atlantic silversides, spot—in some years, windowpane, scup, and northern sea robin.

Notes from the Road—A Developing Fishery for Toadfish

The oyster toadfish, or oyster cracker, is a relatively small, not very abundant fish, that lives in the salty water of estuaries (see below). These fish are unique because they have among the largest (0.2 inches) eggs that are attached to various surfaces for development. They are also distinctive because they undergo development, after hatching, but remain attached to the egg attachment site. The juveniles and adults prefer structured habitats such as eelgrass beds and oyster shells. Despite their name, they do not feed on oysters. In fact, they feed on crabs that feed on small oysters; thus, they actually support oyster growth and survival.

Although this species is not commercially viable in the traditional sense, several minor uses have developed in the last several years. They have always been available, to some degree, from blue crab pot catches from spring through fall. This bycatch allows for some unusual uses, and among the most common, served as live fish in restaurants. These dining spots typically feature a large aquarium from which a customer can choose their dinner. The offerings might include various fish, but toadfish are more common in recent years. They taste like fish, based on our periodic hook and line fishing expeditions, in which we eat whatever we catch. Another use is in hard clam culture operations. After placing the seed clams on the bottom, small-mesh nets cover them to keep common predators away such as blue crabs and mud crabs. Some researchers have observed that releasing toadfish under the nets allows them to eat the crabs and thus prevents the small clams from being eaten. Both uses of toadfish will not have a significant effect on toadfish numbers, but could offer a small, but sustainable, fishery for this unlikely species.

Oyster toadfish with a large mouth and teeth, smooth skin, and dark coloration that characterizes this fish.

Large juvenile and adult fish carrying acoustic tags provide the clearest examples of bay and channel use. As a result, they can be tracked with stationary receivers, not unlike an E-Z Pass system for automobiles, or from a boat. These extensive studies produced results for selected species: summer flounder, hickory shad, bluefish, weakfish and striped bass. In general, these species reside in the estuary, including the deeper waters behind the inlet, during the summer and fall. They all leave by December and migrate through the inlet as they begin coastal migrations southward and then return in the spring. These seasonal migrations relate to reproduction in estuaries, including those for weakfish and striped bass.

This same tracking approach also helps define smaller-scale habitat use patterns in these deep waters. For example, we tracked a representative striped bass (named "Nicky") tagged in Shooting Thorofare during June and July 2004 (Fig. 8.28). During both periods, it remained resident off The Point at the end of the Sheepshead Meadows, during the day and night. This fish, however, made much more extensive movements to the mouth of Little Sheepshead Creek during the night of June 15, perhaps seeking to feed. Representative of many tagged striped bass tracked in the system, this fish remained resident for long periods during the summer in deep but nearshore areas but also left the system for the winter. It returned to this estuary and the same location the following year.

Marsh Thorofares

The thorofares through the Sheepshead Meadows and the adjacent islands provide another link with the inlet and the estuarine system. Since they offer easy and reliable access to the waters flowing underneath, the bridges on Seven Bridges Road provide the best identification and understanding of which fauna use these links. As a result, we at the Rutgers University Marine Field Station, and others, have taken advantage of that access over decades to "read" the waters using a small mesh (one twenty-fifth of an inch) plankton net at the bridge over Little Sheepshead and other creeks to sample the larvae and small juveniles of animals in the water column on flood tides. This sampling method determines those present and, further, those that are moving from the ocean into Little Egg Harbor and Great Bay. We sample at night to reduce avoidance of the nets (see Notes from the Road: Faunal Sampling Techniques, in Chapter 6). Our primary interest lies with the fish, and we have characterized their annual and seasonal occurrence over three decades using weekly sampling.

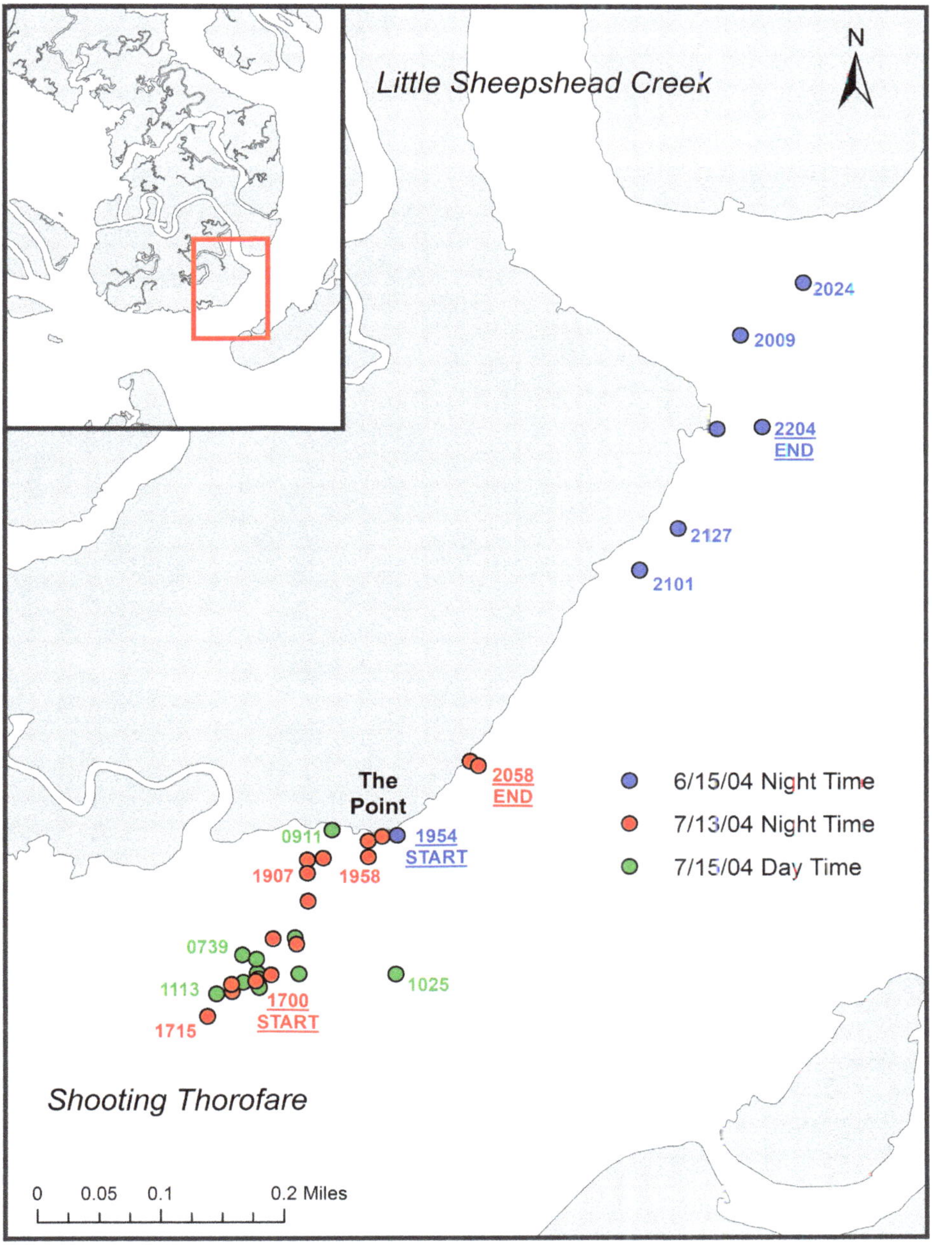

Figure 8.28 The movements of an acoustically tagged adult striped bass (named Nicky, 28 inches) during day vs. night tracking from a boat during June and July 2004 at the end of the Sheepshead Meadows near Little Egg Inlet. "Start" and "End" indicate the beginning and end of each tracking sequence on a given date.

The early life history stages of several fish species are common, or even dominant, depending on the year and the season (Fig. 8.29). The fish larvae and juveniles that spawned a long distance away include those from the east coast of Florida (gray snapper, ladyfish), North Carolina

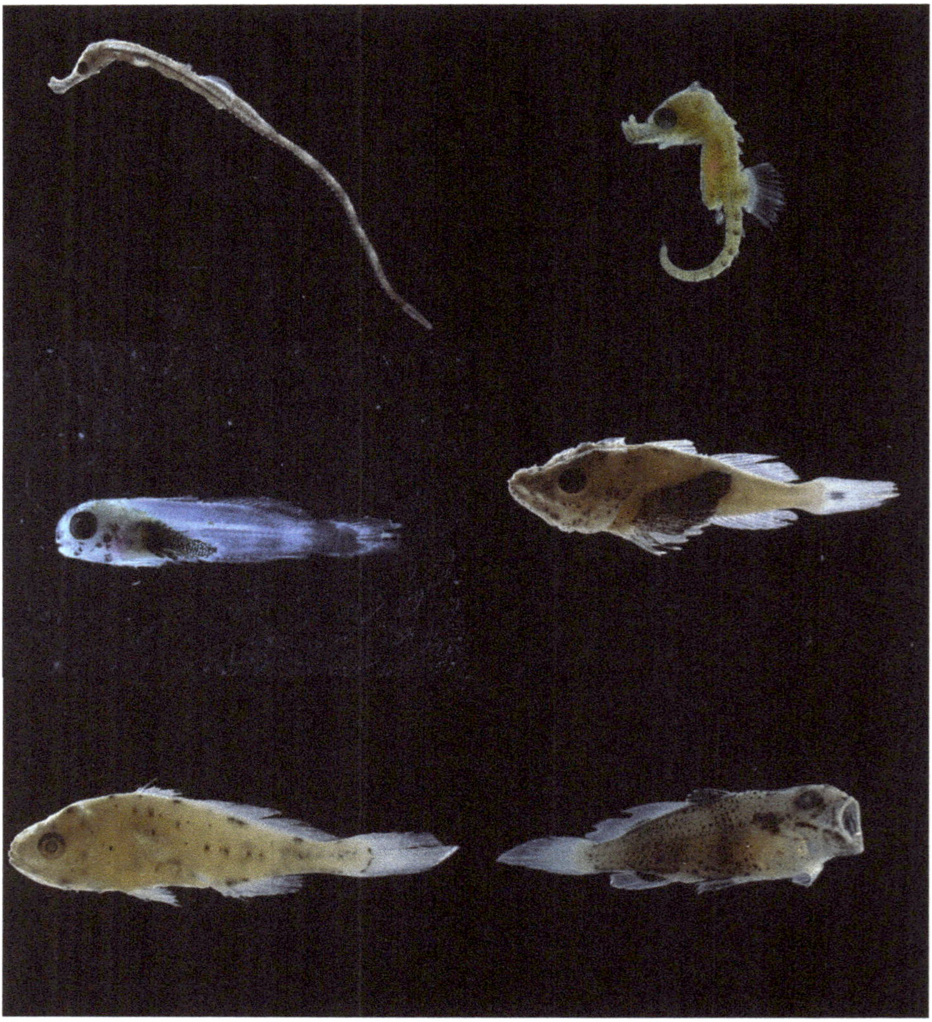

Figure 8.29 Representative early life history stages of typical fishes found around the Sheepshead Meadows including the larvae of (top left) northern pipefish, (top right) lined sea horse, (middle left) feather blenny, (middle right) striped sea robin, (bottom left) Atlantic croaker, and (bottom right) northern kingfish.

(striped burrfish, jack crevalle, spotfin butterflyfish, goatfish, lookdown) (Fig. 8.30), and Massachusetts (pollock, Atlantic cod). The American eel demonstrates the clearest link between the distant ocean and the estuary. We have monitored glass eels and elvers of American eels (Fig. 8.31), which occur throughout the Sheepshead Meadows, from the bridge over Little Sheepshead Creek and in the Grassle Pocket Marsh. They are particularly interesting because they travel the greatest distance of any fish to reach the area. These eels spawn in the Sargasso Sea and then travel the waters over

Figure 8.30 Representative fishes from spawning south of Cape Hatteras that have been collected as juveniles around the Sheepshead Meadows including (top left) burrfish, (top right) permit, (middle left) goatfish, (middle right) cobia, (bottom left) spotfin butterflyfish and (bottom right) lookdown.

Figure 8.31 Glass eels (bottom) and elvers (top) stages of the American eels that are commonly collected in the Sheepshead Meadows.

the East Coast's continental shelf. They enter Little Egg Inlet and Little Egg Harbor, and travel up several thorofares through the peninsula, including up Tuckerton Creek, to reach Tuckers Creek in the Grassle Pocket Marsh. The eels also move even further up the Mullica River. This extended movement from spawning grounds to our estuary is an incredible journey, but it occurs annually.

All or most eels take advantage of tidal stream transport, i.e., they take advantage of flooding tides to get closer or through the inlet but may drop to the bottom or hug the shoreline during ebb tides, thereby retaining their position until the next flood tide. Such actions provide the rationale for why we sample behind Little Egg Inlet at Little Sheepshead Creek on flood tides. This can happen during all seasons, but with low temperatures, and especially near freezing, they may remain on the bottom. During these periods, the fish larvae entering through the inlet (summer flounder, American eel) are much reduced in our collections. Seasonal occurrence varies between species with different species arriving from the ocean in the winter (Atlantic herring, spot), spring (windowpane, grubby, American sand lance), summer (speckled worm eel, Atlantic menhaden, striped anchovy, black sea bass), and fall (Atlantic croaker, summer flounder). Other fish, resulting from local reproduction in the estuary, appear in the spring (winter flounder), summer (striped searobin, northern pipefish, fourspine stickleback, Atlantic silverside, lined seahorse, silver perch, naked goby), and fall (darter goby).

Sampling from the bridge over Little Sheepshead Creek provides the primary data for understanding these patterns, but we have made comparisons from the bridges over Jimmies and Little Thorofare creeks in the Sheepshead Meadows and these are very similar in the species captured and their seasonality. We have also compared identical sampling from Barnegat Bay, including behind Barnegat Inlet and the Point Pleasant Canal, and found similar patterns. Thus, the Sheepshead Creek samples represent much of southern New Jersey inlets and the thorofares behind them.

The bridge over Little Sheepshead Creek has served a central role for observing fish passing through the thorofares especially relative to reproduction. Since many fish create sounds during courtship and reproduction, we have recorded these species-specific sounds using underwater microphones deployed in the creek. These studies have identified several species that reproduce in the Sheepshead Meadows vicinity, including oyster toadfish, weakfish, and cusk eels, and recorded the same species at a site just outside Little Egg Inlet (Fig. 6.7). Yet another example of fish using the thorofares has occurred at Little Sheepshead Creek, where we have consistently detected acoustically tagged weakfish, perhaps because it is the Sheepshead Meadows' deepest thorofare. It is certainly a focus of weakfish fishing activity (Appendix 2).

Both juvenile and adult fish appeared in trawl samples from the Little Sheepshead Creek thorofare where the bottom had many substrate types and somewhat different species dominated (black sea bass, spotted hake, hogchoker, spot—some years).

Marsh Surface and Creeks

A variety of killifish, including mummichog, sheepshead minnow, rainwater killifish, and inland silversides, dominate the Sheepshead Meadows marsh pools and the adjacent islands. Other species there include the blue crab and grass shrimp species (Appendix Table 1), based on extensive sampling over the last several decades. All these fish lay attached eggs, which ensure that the young stages remain in these habitats.

The marsh creeks of the Sheepshead Meadows, and the adjacent islands, contain skinny or shallow inshore waters used as nurseries for a large number of fish species. We know this based on the extensive studies conducted in the Sheepshead Meadows's high salinity waters. My previous book, *Beneath the Surface*, partially documented these studies. Additional marsh creek studies at high and low tides during the day and night over several years provided additional insights into intertidal and subtidal transients. We sampled intertidal creeks with weirs and seines and found small species of fish (Sheepshead minnow, inland silverside). Fish and crabs that use intertidal creeks are the source of the fish and crabs that use the marsh surface on high tides. We built a flume/camera system (Chapter 6) for filming their occurrence and behavior during September along Willit Creek, adjacent to the Rutgers University Marine Field Station. Mummichogs, Atlantic silversides, and gobies (probably naked goby), and green crab and blue crab all dominated the photo-documentation. The mummichogs demonstrated the clearest behavior at high tide by feeding on algae mats at the saltmarsh cordgrass base. These same individuals might have used the algae mats in creeks to avoid predators such as summer flounder and bluefish, both common in Willit Creek on high tides, based on other sampling and acoustic tracking. The film recorded all the above-named species leaving the marsh surface on the falling tide. Other fish appear in both intertidal and subtidal creeks (mummichog, striped killifish, Atlantic silversides, three-spine stickleback, white mullet, naked goby). Those most abundant in the subtidal creeks average relatively larger in size (smooth dogfish, American eel, Atlantic menhaden, Atlantic herring, bay anchovy, pollock, Atlantic needlefish, northern pipefish, bluefish, spot, northern sennet, summer flounder, and winter flounder).

Other marsh habitats include naturalized embayments such as the Rutgers University Marine Field Station boat basin, where extensive tag/recapture studies have clarified habitat use pattern for small juvenile fish. Using individual coded wire tags and recapture techniques (Chapter 6) clearly indicates that some species remain resident for extended time periods during the summer and fall. We have recaptured some individual tautog thirteen times and individual cunner up to six times. Other studies, using the same techniques, documented similar patterns for black sea bass and spotfin butterflyfish. In all these studies, individual growth was typical of that from other habitats.

The movement scale varies with the species that live or migrate through these Sheepshead Meadows habitats. It can often be small scale (resident) for small species, such as mummichogs, and larger (throughout the estuary) for others, such as fast growing, larger juvenile bluefish. Others, such as bay anchovies retreat to south of Cape Hatteras from New Jersey and the Sheepshead Meadows for wintering.

Some of these fish also use human-altered habitats that occur along Seven Bridges Road. One of the most common is the Grassle Pocket Marsh near the junction of this road with Oak Lane at the border between Tuckerton and Little Egg Harbor Township. The variable salinity of Tuckers Creek influences the fish composition. Observations using a continuously recording submerged data logger, which the Jacques Cousteau National Estuarine Research Reserve provided, and observations at the time of collections with a salinity meter (refractometer), clearly indicate that the salinity is very variable with values ranging from 0–21 parts per thousand. The highest salinities occur at the creek's lower end due to the downstream Tuckers Creek-Tuckerton Creek confluence and eventually joining Little Egg Harbor and the ocean through Little Egg Inlet (Fig. 7.25). The upper portion of Tuckers Creek typically has fresh water as the result of these waters coming from streams through two wooded areas that connect with the marsh by culverts under Radio Road.

At this location, fed upstream and downstream through culverts under roads (Chapter 7), we have attempted to characterize the fish species that use this marsh. The salinity level influences individual species abundance. We only collected the eastern mudminnow at the creek's upper freshwater end while the mummichog and banded killifish were most abundant in traps at the creek's lower and middle portion, respectively, primarily in the summer. Bluegills were most abundant in the creek's middle and upper portions. Other species (sheepshead minnow, fourspine stickleback) were rare.

The most unusual species collected are the glass eel stages of the American eel that have recently arrived from the distant Sargasso Sea. The larger mesh traps did not detect these eels, but they appeared in eel collectors (weighted mats of fine-mesh unraveled rope) deployed during the spring. At Holly Lake, now a totally freshwater lake, juveniles of typical freshwater species (sunfish—pumpkinseed and bluegills, largemouth bass) and yellow eel stage of American eel that passed over the lake's dam, probably as glass eels, dominate collections in the same trap type.

Estuarine Beaches

The sandy beaches along the water's edge in the Sheepshead Meadows can feature a variety of faunal types using the beaches for reproduction, feeding, burial and predator avoidance. The value and faunal types using them can vary with location. The striped killifish is among the most frequent and abundant fish on sandy beaches from Little Egg Inlet to the Sheepshead Meadows' northern end. This species uses these beaches in a variety of ways, all of which account for its usual restriction to this habitat type. We followed many aspects of this species' life history and ecology at a sandy area immediately adjacent to the Big Thorofare oxbow at the First Bridge along Seven Bridges Road. Feeding on these beaches occurs when the juveniles and adults probe their snouts into the sand to gather invertebrate prey. Reproduction also occurs there when clasping males and females bury the fertilized eggs in the sand, where the eggs remain until hatching. The timing of reproduction, based on the female's gonads weight, peaked in May–July in 2015 and May–June in 2014. After hatching, schools of small individuals (about one half inch) can be seen swimming in 1–3 inches of water over the sand, where they move with the tides and ride small waves in these shallow waters. Based on laboratory observations, we know that even the small juveniles (about one half to one and a half inches) can bury in the sand, thus their whole life history remains tied to sandy habitats.

Other common faunal components of the Sheepshead Meadows taking advantage of these sandy beaches include horseshoe crabs, lady crabs, blue crabs, and sand shrimp. The horseshoe crabs use the beaches as spawning sites and nurseries for the buried eggs. Once they have hatched, the larvae can be found in the plankton but then settle into shallow subtidal sandy sites to feed and molt, growing for as long as several years before migrating into the ocean, only to return when they are ready to spawn. The sand shrimp, blue crabs, and lady crabs can bury completely out of sight on these sandy beaches, potentially to avoid predators, but the lady crab is especially adapted to this behavior. Based on our laboratory observations, they can bury deeper than blue crabs and can do this by reversing the flow over their gills.

Chapter 9
Reading the Road

Introduction

The Sheepshead Meadows is an ideal location for natural history studies, and, more broadly, the "ecology of place," i.e., the relationship between the plants and animals and their environment. This is particularly true of relatively unaltered landscapes and habitats as they provide a baseline for understanding how natural systems work and their restoration if altered. The Sheepshead Meadows is such a place, and it has the advantage, in some respects, because nearly six miles of Seven Bridges Road bisects the marshes. This road, and its bridges, provide visual access to many meadow parts during all seasons of the year, both day and night, and has done so over decades. As a result, prior studies and observations of the road and its surroundings have provided terrestrial and aquatic habitat descriptions (see Chapter 7) and flora and fauna (Chapter 8) for these productive landscapes. Further, even when tides and storms limit access, these flooding events provide important insights into these increasing effects associated with climate change and sea level rise (Chapter 10). In addition, human influence in the marshes along today's Seven Bridges Road, from precontact to the present, has altered the Sheepshead Meadows and provided insights into the future.

Landmarks from the Road

The flat, low elevation terrain of the salt marshes along Seven Bridges Road, especially after leaving the slightly higher land around Tuckerton, allows many local landmarks to be seen at some distance. The high elevation at First Bridge, over Big Thorofare, provides the possibility of observing the buildings along Radio Road to the south, with the tallest buildings associated with the former wireless tower. Further south, on exceptionally clear days, Atlantic City buildings may be visible. In the opposite direction, on the other side of the bridge, the red, white, and blue water tower in Tuckerton stands out (Fig. 9.1). From the same vantage point many low relief islands that comprise the Story Island complex in Great Egg Harbor are visible. Behind them, further to the east, you might be able to see the water towers at Beach Haven and elsewhere on Long Beach Island.

Other bridges along the road, albeit lower in elevation, provide good views of local landmarks because vegetation does not block the vista. For example, the bridges at Little Thorofare, Jimmies Creek, and Big Sheepshead Creek, including much of the road in between, provide views on clear days of the gunning club buildings on Drag and Hither islands to the north, if you know where to look. The water towers on Long Beach Island are also visible, except when there is fog, especially as you get closer to the road's end. On the road's opposite side, and over the same area, the ruins of the fish factory on Crab Island are easily visible. By the time you reach the bridges over Jimmies Creek and Big and Little Sheepshead creeks, the red roofs and white buildings of the Rutgers University Marine Field Station are obvious to the south. Also, just before Little Sheepshead Creek stands the meteorological station associated with the Rutgers University Department of Marine and Coastal Sciences. The equipment occupies the former platform used as the Crab Island Fish Factory weigh station/warehouse. At the road's terminus, you will find a sandy path leads to Shooting Thorofare (Fig. 9.2). From there, keen observers will view the land mass and beaches

Figure 9.1 Aerial view from a helicopter of the prominent red, white, and blue water tower along Green Street in Tuckerton with Tuckerton Creek winding past marshes and marinas as it flows toward Little Egg Harbor.

at Little Beach to the south, the Holgate portion of the Forsythe National Wildlife Refuge, with Little Egg Inlet between. This view reveals abundant open water with fast tidal currents which provides potential insight of the state's reasoning for not completing the final portion of the Seven Bridges Road to Brigantine and Atlantic City. These landmarks are best observed on clear, bright days. They are less visible on foggy days, during storms, and in winter due to the shorter periods of daylight. At the road's mainland end, the historic town of Tuckerton contains the Tuckerton Seaport and the Tuckerton Historical Society, both located off Route 9. These institutions provide access to the area's cultural history.

Reading the Road at Three Speeds and Four Seasons

Our perception of events along the road varies with our speed of travel, the season, and the frequency one travels the road. In my case, I estimate my trips over Seven Bridges Road total over 18,000 times, in all seasons, stages of the tide, during storms, etc., during my 33-plus years working at the Rutgers University Marine Field Station. Automobile traffic on the road typically moves at slow speeds, resulting in several speeds (driving, biking, walking, staring while stopped) being available. (Although traffic volume clearly increased during Covid 19's worst days due to people's desire

Figure 9.2 View of the open waters of Shooting Thorofare from a sandy path from the very end of Seven Bridges Road with Little Beach on the far horizon.

to get outside.) All these approaches can be thought provoking, especially when observing vegetation changes and fauna behavior along the road. The fauna changes with the seasons, especially the birds, from waterfowl in the winter to the numerous kinds of herons and egrets, and the variety that pass through during their north–south migrations along this portion of the Atlantic Flyway (Chapter 8). The speed you travel is central to what you can see. If you are out sightseeing in a moving car, you can grasp general differences in roadside habitats and how they change with the seasons. The vegetation is primarily perennial and therefore consistently present across seasons but varies in color with seasonal flowering. The marsh colors change from green in the spring through summer, then turning golden during the fall and eventually the marsh wears winter brown. These are hard to miss.

The vegetation reflects the road elevation (Chapter 7). The adjacent intertidal marsh vegetation and the roadside vegetation display, throughout the road's length, subtle variations due to small-scale changes in the elevation. The invasive common reed, which varies in occurrence, abundance, and height, usually in response to the higher salinity waters flooding from the adjacent bays, demonstrates the roadway changes. Thus, the road's low places feature no common reed, and few or no shrubs; natural marsh grasses proliferate there (Fig. 9.3). At higher elevations, such as at the approaches to the bridges (Fig. 7.19), the common reed is tall and abundant, but the

Figures 9.3–9.4 (Previous page) The lowest portions of the road are frequently flooded with high salinity water and this is reflected by the absence or smaller size of the invasive common reed which can be replaced by saltmarsh grasses. (Right) The highest elevations along the road are at the approaches to the bridges and that is where the tall common reed is dominant.

transition from these two extremes is observable over a short distance (Fig. 9.4). In other words, you can detect even subtle roadway elevation changes and its right-of-way by paying attention to the vegetation.

The differences in the elevation along the road also support different faunal species, so, the slower you drive, the more likely you are to detect these. This applies to slow-moving fauna such as the diamondback terrapins, which are often spotted digging nests. The three marinas and the state-owned landings along the road provide ramps for motorboat, canoe, or kayak launch and recovery. Moving slowly provides rich opportunities to observe bird fauna. While the paved road itself does not provide habitat, some birds actually feed on the road. These include redwing blackbirds, attracted to insects that gather, or are more visible, on the road. Other times, boat-tailed grackles can be seen picking at clam and mussel remains gulls have dropped. Clapper rails, sometimes with a line of chicks, occasionally can be viewed crossing the road, especially during very high tides. Other birds, such as turkey vultures, and in recent years, black vultures, feed on the occasional mammal roadkill. In other instances, different shorebirds appear to feed on the road when flooded. Numerous other birds, especially those that feed in the marshes, can be quite visible, as the frequent birders have observed (Chapter 8). Those commonly visible at low travel speeds include male redwing blackbirds setting up territories and building nests in the tall vegetation adjoining the roadway. Herons and egrets will occupy the road's lee side marsh vegetation, especially during wind events. In these situations, the same birds may roost in the tall trees at the bridge approaches (Fig. 9.5).

Remaining stationary provides the most appropriate way to observe faunal behavior. This approach can work especially well while sitting in your car because it serves as a "blind" and is less frightening to the birds. Remaining still, I have watched juvenile American bittern stalking fish prey in shallow pools; the adults are especially difficult to spot as they remain motionless in nearby vegetation. The bridges serve as ideal vantage points for other reasons. You can often gain a good sense of the thorofares' tidal stage under the bridges but be aware that the flow direction varies with location. The tide floods from the same direction at Little Sheepshead Creek

Figure 9.5 Herons and egrets are temporary residents in the taller trees at the higher elevations near the bridges.

(from Little Egg Harbor towards Great Bay), but in the opposite direction under the four bridges closest to Tuckerton, including Big Sheepshead Creek, Jimmies Creek, Little Thorofare and Big Thorofare. Of course, strong winds can affect the flow's direction. The thorofares and their ice cover reflect winter water temperatures, but the ice varies with location. For example, the bridge over Little Sheepshead Creek is often the last place to feature ice cover on any of the thorofares due to its proximity to warmer ocean waters, and its deeper depth. Ice-covered thorofares more likely occur further from the ocean. The second and third bridges from Tuckerton, over Jimmies Creek and Little Thorofare, are crabbing and fishing hotspots, and their catches demonstrate the current faunal abundance. The shallow edges under the bridges, especially the bridge over Little Sheepshead Creek, are good observation places to see spawning horseshoe crabs during the early summer. The large number of laughing gulls congregating at these sites indicates the horseshoe crabs' presence because they are feeding on the horseshoe crab eggs. Don't stop your vehicle on this bridge and that over Big Sheepshead Creek, however, because stop lights regulate the traffic on these narrow, single-lane bridges.

Seasonal changes along Seven Bridges Road provide one of the most striking and interesting experiences. These transitions occur, in large part, as a function of air and water temperature changes, and shortening day length. In turn, the seasonal occurrence influences the flora and fauna, above and below the water, that can be observed. The seasonal fauna includes insects, birds, and mammals but also those lesser-known underwater animals and plants like macroalgae, jellyfish, shrimp, crabs, and fish. Various roadside flora species respond to other changes, including precipitation amounts, and water salinity levels.

Of course, traveling on Seven Bridges Road at night provides you with yet another perspective. For example, flooding frequency and location can be greater because, in general, the tides run higher at night. It is also more difficult to see nocturnal animals, including diamondback terrapins, red foxes, raccoons, and mink, accounting for some roadkills on the road.

Reading the Road from the Pocket Marsh

The rare opportunity to observe from within the marsh center is possible at the Grassle Pocket Marsh. The entrance is located at the Jacques Cousteau National Estuarine Research Reserve parking lot. After a short walk on a distinct path through a maritime forest that has replaced the former farm at this site, you can walk out onto a wide pier that leads right into this pocket marsh's center and ends at Tuckers Creek. From there, you can observe the marsh without boots, and the adjacent upland forest. This vantage point clearly shows the invasive common reed that dominates the former salt marsh, especially while looking upstream. Downstream, towards Seven Bridges Road, the remnant salt marsh becomes more visible. Chapter 10 details this common reed transition due to rising sea levels. The Rutgers University Marine Field Station Open House (typically on a Saturday in late September, Chapter 4) provides another Sheepshead Meadows salt marsh access point without boots. The nearly one-quarter-mile causeway leading to the Station provides excellent views of the marsh surface, meandering creeks, and marsh pools at all tidal stages.

Pattern of Road Flooding

The roadway's elevation prevents flooding on an average day. On spring tides and during storms, however, flooding can be frequent and enduring. Some road sections feature variable micro-elevation changes which make the flooding more obvious (Fig. 9.6). Flooding occurs less frequently at the two ends of the road, before First Bridge over Big Thorofare, and after the bridge over Little Sheepshead Creek, because these roadway sections, constructed first, feature higher elevations than the intervening sections. The red cedars prominent along the roadway demonstrate the elevation changes, along with the variety of shrubs found throughout the sandy portion where the road ends just before Shooting Thorofare. Further, there is a general match between the areas of more frequent flooding and the roadside vegetation with edges of regular salt- marsh grasses (saltmarsh cordgrass and salt hay) (Fig. 9.6). This area extends between the

Figure 9.6 Saltmarsh cordgrass along low elevation portion of Seven Bridges Road. Note location of the Rutgers University Marine Field Station in distance.

bridge over Little Thorofare and the bridge over Little Sheepshead Creek. The lowest elevations become evident the morning after a high night-time tide, when the road is still wet from the flooding.

Transport of Marsh Production

One definitive way to read the road is by paying attention to the distribution and amount of wrack, or accumulated dead/decaying vegetation, on the road. The wrack here is consistently composed of dead marsh grasses, particularly saltmarsh cordgrass, and common reed. Its occurrence is episodic and typically related to storm tides and wind that causes the dead vegetation to be rafted from the marsh surface, where it can collect into windrows, where it frequently concentrates along the margin of the road due to the higher elevations of the road and the taller roadside vegetation. These Northeast storms, occurring primarily during winter, wash and blow through the Sheepshead Meadows peninsula's thorofares (Fig. 9.7), accounting for the biggest wrack collections deposited along the north and east roadway edges and upon the road itself (Fig. 9.8). Where this wrack consistently collects on the marsh, it can smother the vegetation and expose the underlying marsh peat. The lowest elevated road sections that experience frequent flooding feature typical marsh vegetation up to the road's blacktop edge. Wrack seldom accumulates on the portions of the road that are more elevated and lined with trees, such as that road section that travels through the marsh near Tuckerton. Accumulation of wrack over time is one index of marsh plant (detrital) production which contributes to marsh sediments and feeds the base of the food chain supporting invertebrates and fishes throughout the estuary. Wrack also creates habitat on the marsh surface for invertebrates (amphipods, isopods and other crustaceans) that provide hatchling diamondback terrapins with a food source. The terrapins prefer burying in and under the wrack. Wintertime brings ice-rafted marsh edge (peat reef) that can be carried onto the marsh surface and be observed there over months until the marsh peat fragments and disappears.

Storm-driven high tides often deposit human debris (e.g. docks, walkways, timbers, and the occasional boat) on the road. The dock remnants and their assembly nails can flatten tires, especially in a storm's aftermath when the wrack covers them, making them difficult to see. These random flotsam pieces likely originate along the bayside of Long Beach Island due to the prevailing northeast winds during most storms. Other trash sources include the occasional black trash bags and other material that appear along

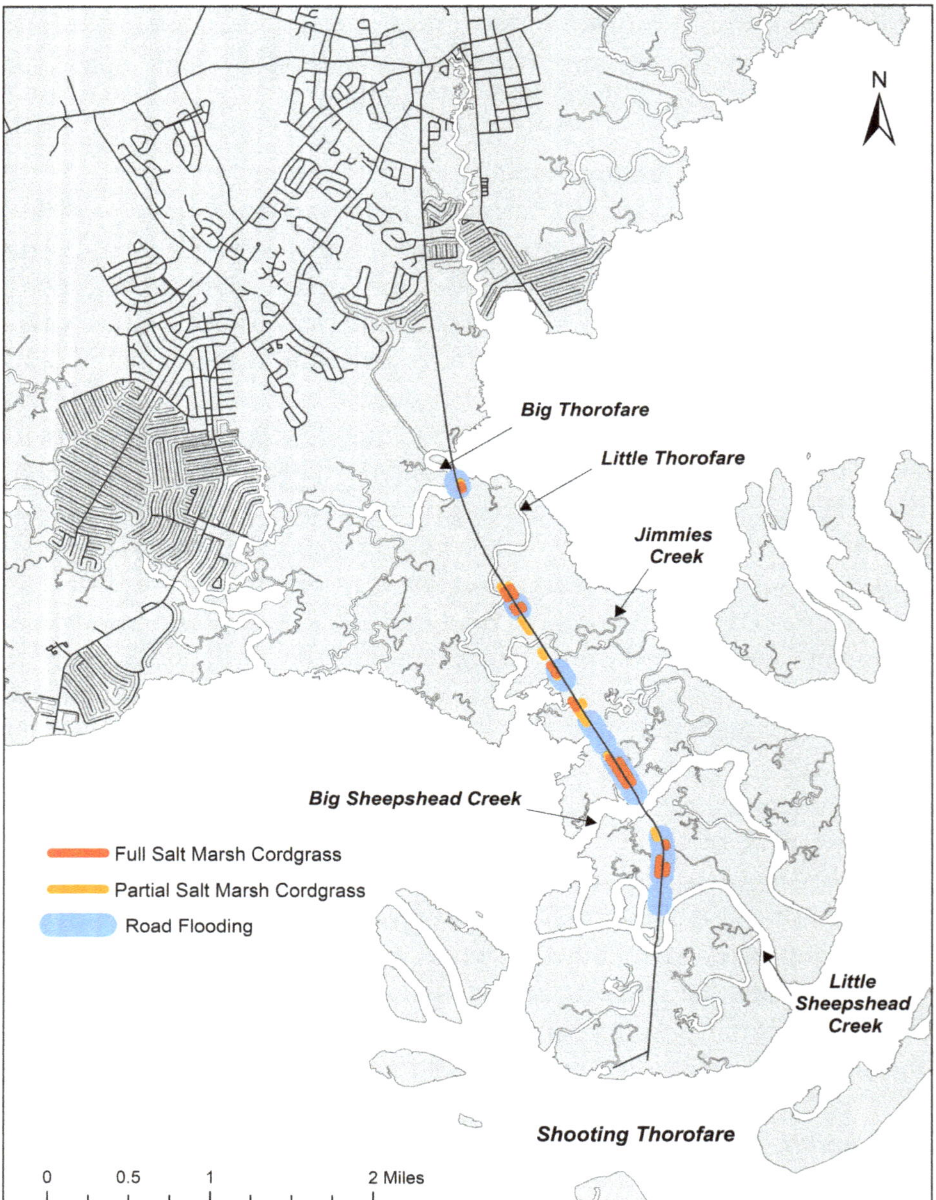

Figure 9.7 Map of frequently flooded road areas that reflect low elevation. These seldom occur at the higher elevations leading to bridges. Prepared by the author and Ryan Larum.

the road on certain dates. These fortunately can be sourced to local groups (e.g. Southern Ocean Birding Group, AmeriCorps Watershed Ambassador Program, and the Barnegat Bay Master Naturalists as part of the Barnegat Bay Partnership Blitz) that periodically conduct clean ups along the road. Little Egg Harbor Township Public Works pick up the bags. Those that frequently travel the road appreciate these cleaning activities.

The Road Facilitates Gull Feeding on Shellfish

The elevated surface of Seven Bridges Road provides a long, hard surface that stretches from Tuckerton to Shooting Thorofare, but juvenile and adult herring gulls only use selected sections when they are feeding on shellfish. They collect shellfish, either single intertidal/subtidal hard clams or sometime clumps of intertidal, smaller ribbed mussels, in the adjacent marsh waterways and drop them on the roadway from a sufficient height to break the shells and feed on the meat inside. These shellfish drop zones are typically adjacent to marsh creeks or thorofares, but their exact location differs along the road. The most prominent areas are near Big and Little Sheepshead Creek and extending to Shooting Thorofare (Fig. 9.9). This behavior only occurs in winter, apparently because other food sources are unavailable. This is when freshly dropped shells remain visible on the road, but the remnant, crushed shells lay on the roadsides year-round (Fig. 9.10).

Figure 9.8 Accumulated wrack near the end of Seven Bridges Road after a storm where it has collected against the taller common reed vegetation. Note the Rutgers University Marine Field Station in the background (upper right).

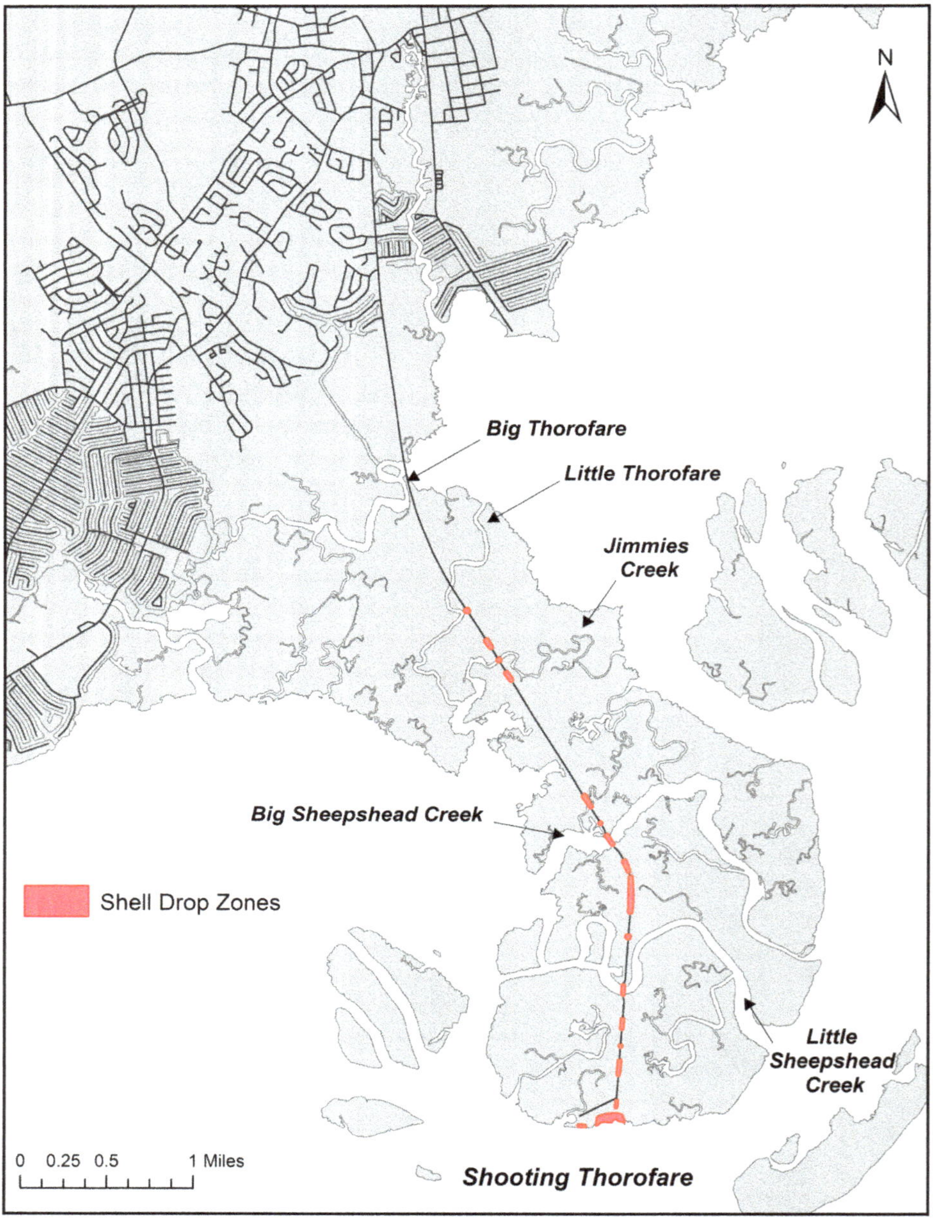

Figure 9.9 Shellfish drop zones, typically hard clams and ribbed mussel, along Seven Bridges Road based on multiple observations during the winter over decades. Prepared by author and Ryan Larum.

The Road as Nesting Area for Terrapins

The roadway's elevation provides nesting possibilities for diamondback terrapins that nest above high tide line and thus protects the nests from tidal flooding. This elevation factor is further accented at the five bridge approaches that feature a much higher elevation than the roadway itself. Multiple individuals from multiple institutions conducted extensive research over many years along Seven Bridges Road, and in nearby Barnegat Bay, providing the data for the following summary. The "Turtle Garden," the manmade small "dune-like" elevation at the former Rand's Marina's west end (Fig. 1.7, 7.30), where abundant nests can be found, proves how important elevation is to the nesting terrapins. These elevated, sandy sites are optimal, especially during late-May through July, when the females can be very abundant, so much so that it requires driving around them to get down the road. As examples, in one drive on the road in recent years, from the beginning of the road in the salt marsh to the end we encountered 53, and in another instance 87 females. During the first drive, repeated a few hours

Figure 9.10 Sides of the road with crushed shells in a herring gull drop zone. Their remains can be seen as shadows along the sides of the road surface during all seasons.

later on the same day, we found less than 10, providing a good indication of how abundant they can be, but also how quickly they can leave the area after egg laying (Fig. 9.11). It is also apparent that some females will revisit the same sites (within 160 feet of where they were first tagged) multiple times in the same summer to lay eggs, based on recapturing tagged individuals. The roadway's elevation has one disadvantage: thick common reed growing in long linear patches along the road probably prevents females from traveling between the marsh and the road.

Driving the road fairly frequently during early summer, observant people will learn that the timing of the female terrapins appears to coincide with the higher new and spring tides. At that time, it is easier for them to swim to the road than walk through the dense marsh surface vegetation. The terrapin abundance causes frequent road mortalities. Such losses among the females are especially critical. A dead female not only prevents individual egg laying again later in the season, but it prevents her from doing so in future years.

Concern about the impacts of road mortality has led the Conserve Wildlife Federation of New Jersey to attempt a number of measures to reduce this mortality. Ben Wurst and colleagues erected a low-level barrier along much of the roadside of the road between the First Bridge and Tuckerton to prevent diamondbacks from getting on the paved road. But you might miss this barrier if you are not paying attention, especially during the summer when the vegetation grows high. They have also erected warning signs along the road with information describing when the terrapins might be on the road. These measures attempt to make the driving public aware that they might encounter this threatened species. These measures appear to be working and reducing the female road mortality over the three decades that I have driven the road multiple times a week.

The recently hatched juveniles (Fig. 9.12) present a generally unrecognized source of road mortality, as they cross the road, due to their small size and color, similar to the road's blacktop surface, thus preventing them from being visible to drivers. This may be particularly important near the Turtle Garden, where nests are abundant and also at the bridge approaches as confirmed in earlier studies. These small individuals appear after the nesting season, usually in late summer (September) from summer egg deposition or during the early spring (April) after they have overwintered in the nests. The importance of nest elevation relative to sea level rise is critical to their long-term survival with some projections for Chesapeake

Figures 9.11–9.12 (Top) Diamondback terrapin digging nest along the side of Seven Bridges Road. (Bottom) Just recently hatched diamondback terrapin as observed at the Turtle Garden.

Bay indicating a 25–55% nesting site loss in the next ten years and over 80% by the end of the century. Terrestrial predators such as birds, red foxes, and raccoons are other mortality sources. These predators' actions are most noticeable because they dig up and eat the eggs immediately and leave the shells scattered near the nest's surface. Both herring and laughing gulls also prey upon the recently hatched terrapins. At the Turtle Garden, these gulls can be seen on poles or the fence surrounding the garden during hatching season. Once they spot a hatchling, they swoop down and ingest these small, soft-shelled morsels. As an example, while eating lunch in my car adjacent to the Turtle Garden, I observed a laughing gull that had captured a recently hatched diamondback terrapin (Fig. 9.13). As a preventative measure, the Conserve Wildlife personnel install weighted cages over the nest after egg-laying to prevent this predation until they can recover the juveniles and remove them to a safer location.

Figure 9.13 Laughing gull that has just grabbed a diamondback terrapin hatchling from the Turtle Garden, the sparsely vegetated sandy area in the background.

"Reading" Humans on the Road

We share Seven Bridges Road with many other people. Most hail from New Jersey, but I see license plates on vehicles from Pennsylvania, New York, Maryland, Virginia, and Florida. Some are scientists doing marsh research around the road and many others are birders carrying binoculars, large cameras, and tripods. Local area residents often walk their dogs or just enjoy some walking or biking. The roadway also attracts the curious; they want to know if people are fishing or crabbing or they just want an excuse to be outside. Those individuals and groups fishing or hunting from boats are obvious because they leave their vehicles with trailers parked along the roadside. Others launch their kayaks at the ramps along the road for some quality time in a less than crowded area. Some are obviously crabbing from the bridges or roadside creeks or collecting bait (minnows, shrimp) from the roadside creeks and pools. Others appear to be "stormwatchers" because they are seen driving the road, usually in big trucks, during the worst weather. Some are on the road for a specific purpose: to pick up timbers for reuse after a storm or pitchforking the wrack presumably for use in their gardens or landscaping. Several times each year, local organizations provide cleanups on the road to collect the trash that is sometimes left on the road or deposited there during storms. It is also surprising how many vehicles are on the road at night. We note this as we drive the road because our workdays are influenced by tides and field sampling occurs at all hours. Many of those traveling the road do it frequently enough that they know each other and will stop to chat. Some are strictly seasonal, like those who decorate red cedar trees after Thanksgiving for Christmas. Often, they decorate the same tree year after year (Fig. 9.14).

Earlier studies, using traffic counters at several locations along the road to study diamondback terrapin roadkills, found much higher traffic levels on the road closest to Tuckerton. This may occur because some people find the stoplights confusing, seemingly in the middle of nowhere, and turn around. Others are just confused. I have had some people stop me to ask, "Is this the way to Atlantic City?" In the 1920s and 30s, the state envisioned that plan, but fortunately for many of us, that never happened. You can often detect newcomers to the road, especially at the last two bridges' stop lights during the summer, when they breeze up to a red light and suddenly realize their open car windows are open and the greenheads have found them.

Figure 9.14 The frequency with which people drive and enjoy Seven Bridges Road is reflected in the annual decoration of one of the red cedars along the road as a Christmas tree.

Chapter 10
A Changing Environment and Implications for the Future

This chapter addresses the changing environment along Seven Bridges Road in response to a changing climate and the resulting sea-level rise. The chapter also attempts to anticipate the responses of the flora and fauna of the area to these changes, both now, and what may be the future.

Climate Change

New Jersey's location along the East Coast of the United States accounts for the wide seasonal swings in temperature, the strongest environmental variable. As a result, our coastal state typically experiences cold winters, more so than any coastal state to the south, but also warm summers, more so than any state to the north. This variability requires the measurement of long-term temperature changes to ascertain climate effects. Hence, examining annual changes, rather than daily or monthly differences that might simply reflect short term weather effects, will yield a more accurate data source. Based on this source, it is clear that New Jersey's air temperatures have been increasing since 1895, with more frequent warmer years, relative to the long-term average since the 1950s, but especially since 1980 (Fig. 10.1). As a further reflection, 15 of the 20 warmest years in New Jersey have occurred since 2001 and 19 of the 20 warmest since 1990. Of these years, 2023 was the warmest year on record. Overall, average air temperatures have risen more than 3.5°F since the beginning of the twentieth century. The increase in air temperature results in an increase in water temperature, based on long-term measurements taken in the waters around the Sheepshead Meadows. This temperature increase is not just a local phenomenon; the increase in water temperatures reflects a warming Gulf Stream off the East Coast. The Gulf Stream has apparently moved closer to the U.S. coast over the last two decades. In recent decades, ice accumulation has not occurred, as it has in the past, as a result of these warming temperatures (Fig. 10.2).

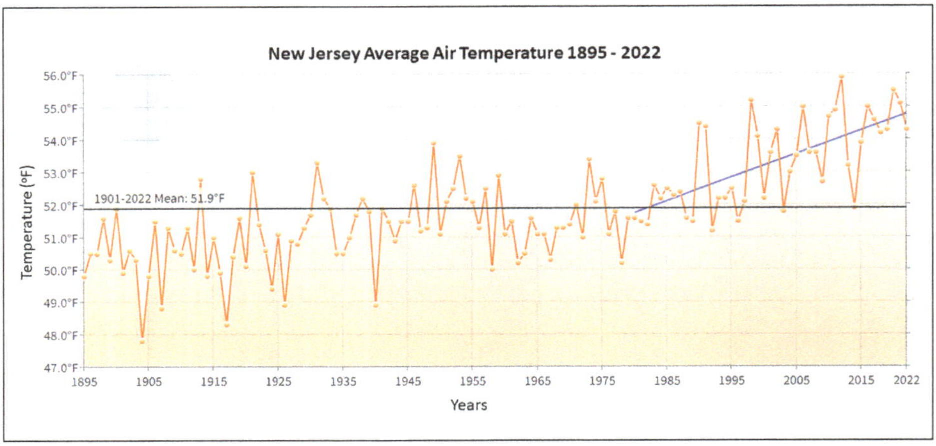

Figure 10.1 Long-term change in New Jersey air temperatures from 1895–2021 indicating increasing rates of change since around 1980. From National Centers for Environmental Information.

Figure 10.2 Accumulated ice flows in the southern portion of the Seven Islands area in Great Bay. Photo of Jim Allen provided by him.

During the same decades, increasing precipitation variability has ensued, with drought in the early 1960s, wetter in the 1970s, and very wet in the last decade. The resultant increase in atmospheric humidity has prompted a trend favoring an increasing frequency and severity of storms. Such low-pressure activity includes nor'easters, which often do damage because their duration frequently continues through several high tides. Other weather disturbances include tropical storms and hurricanes, such as Superstorm Sandy. This was expressed in the Sheepshead Meadows region during Sandy by flooding Mystic Islands where the majority of the 4500 houses in this lagoon development suffered some damage (Fig. 10.3). More specifically, Rand's Marina on Seven Bridges Road experienced complete destruction due to its low elevation (Fig. 10.4). The Rutgers University Marine Field Station, even though it is elevated above the marsh surface, suffered damage when floodwaters ruined the back-up generator, which sat at a lower elevation. Flooding and erosion made the lower part of the causeway, where it meets Seven Bridges Road, impassable. The large timber embedded in the causeway provided further evidence of the flooding as

Figure 10.3 Characteristic damage in the Mystic Island lagoon development from Superstorm Sandy.

Figures 10.4–10.5 (Top left) Damage to buildings at Rand's Marina along Seven Bridges Road due to Superstorm Sandy. Now abandoned, the property is part of the Great Bay Wildlife Management area. (Top right) A large timber embedded in the causeway to Rutgers University Marine Field Station, about nine feet above the marsh, as a result of flooding during Superstorm Sandy. Figures 10.6–10.7 (Bottom left) Destruction of an elevated portion of Seven Bridges Road at First Bridge due to effects of Superstorm Sandy in October 2012. Photo from Ocean County Bridge Department. (Bottom right) The water level from Superstorm Sandy was so high that a Rutgers University Marine Field Station boat was lifted off the highest part of the railway and disappeared. Two weeks later it was found about six miles away up against trees on the mainland in Tuckerton. In this image Station personnel are planning how to refloat the boat.

an example of water height during the storm (Fig. 10.5). Similar intense flooding occurred on Seven Bridges Road at the First Bridge, closest to Tuckerton, where half of the road washed away, making it impassable (Fig. 10.6). The degree of flooding was evident all along the road where boats, docks, and wrack from the surrounding marshes, covered the road or completely washed away (Fig 10.7). This same problem occurred in 1984 with similar damage at the First Bridge (Fig. 10.8).

Figure 10.8 Prior damage to First Bridge over Big Thorofare in the winter of 1984. Image from Kevin Shick.

Changes in the Fish and Crab Fauna in Response to Warming Temperatures

Fish are cold-blooded, that is, they cannot regulate their internal temperatures and therefore are at the mercy of the temperature in the surrounding waters. This makes them sensitive to temperature changes. Collecting seasonal temperature fluctuations regularly and compiling them on an annual basis provides a good indicator of change and thus one of the reasons for this focus. As an example, oceanic water temperature increases along the Northeast Coast of the United States are reflected in the changes in thermal habitat and spawning habitat, and thus, the delivery of larval and juvenile fish to estuaries, including the waters around the Sheepshead Meadows. This is often referred to as a response to climate velocity, i.e., the rate and direction that climate shifts across the seascape.

In response, Great Bay and adjacent estuaries such as Barnegat Bay have experienced a change in the larval fish species composition, based on extensive and intensive studies from 1990 to 2015 (Table 10.1). During that time, a total of 1,252 weekly, nightly samples captured over 650,000 larvae representing 138 species at Little Sheepshead Creek. These larvae arrived in waters coming from the Atlantic Ocean during night flood tides through Little Egg Inlet. This analysis found an increase (18 of 21) in southern fish species larvae (those that originate from spawning in the ocean south of Cape Hatteras). These include Atlantic thread herring, blackcheek tonguefish, darter goby, fat sleeper, green goby, and others (Table 10.1). Over the same period, blue crabs have become more abundant in response to the warmer waters. Further, some of these southern fish species are now established as juveniles and adults in the waters around Seven Bridges Road, where they were absent in prior collections. These include the striped blenny, which did not occur from 1990 to 2006, but has been present since then, and highfin goby, which was not present from 1990 to 2004, but has been present since then. At the other extreme, the estuary has witnessed a decrease (Atlantic herring, inshore sand lance, tautog, cunner) in northern species (those that originate in the ocean from north of Cape Cod) (Table 10.1). Of these, the winter flounder is best studied with an indication that warm, spring-water seasonal temperatures produced fewer larvae and juveniles.

Another way in which warming temperatures influence larval fish in our time-series is through the timing of their occurrence, presumably resulting from a shift in the timing and location of spawning or transport from the ocean through Little Egg Inlet. This outcome stems in part from the warm Gulf Stream flowing closer to the coast than in previous decades, possibly expediting the transport of southern species to the New Jersey coast. A recent analysis of over 25 years of weekly sampling at Little Sheepshead Creek found that four species are occurring earlier (conger eel, weakfish, winter flounder, striped anchovy), while others are occurring later (Atlantic menhaden, summer flounder, American eel). Some of the latter might be explained by an increase in the spawning stock (Atlantic menhaden, summer flounder) or changes in the transport mechanism from the Sargasso Sea for American eel.

Catches of juvenile fish and crabs in traps in the Rutgers University Marine Field Station boat basin, adjacent to Shooting Thorofare, based on collections from 1990 to 2015 (Table 10.1), reflect the increased water temperature. During that time, twice a week collections resulted in over 15,000 samples comprising over 239,000 fish of 32 different species. These fish included some southern species with increased abundance (silver perch), while many northern species declined (cunner, grubby, pollock, three-spined stickleback, winter flounder, rock crab, green crab). These changes can also be observed in commercial fishery species, including the

Atlantic croaker, a southern species, which has increased in recent years and become the basis for a significant New Jersey fishery. This presumably occurred because they can survive during the milder winter water temperatures. Over the time period from the summers of 1998–2013, trawl collections included many juvenile fish during a 3°F increase in average water temperature. Of those caught, five species demonstrated a decline in abundance, including one northern (grubby), two southern (American eel, weakfish), and two local (fourspine stickleback, northern pipefish), while three species showed increasing abundance, including one southern (silver perch), and two local (spotted hake, windowpane) species. While there is suspicion that storms and their increasing frequency might influence the fish fauna of some estuaries, that did not happen in adjacent Barnegat Bay, based on the trawl sampling before and after Superstorm Sandy.

A comparison between larvae and juveniles patterns, including across different sampling gears and different sampling months (year-round for larvae and juveniles in traps, July/September in trawl) may account for the relative inconsistency of some species (Table 10.1). Silver perch was the only species that had the same pattern (increased) across all gears. Other species demonstrated inconsistency, with juvenile spotted hake declining in traps but increasing in trawls. Some species revealed a pattern of increasing abundance for a local species (seaboard goby) and decreasing abundance for other local species (striped killifish, tautog). This general pattern of change in fish species composition extends into adjacent Barnegat Bay. In addition, there is a critical distinction between new species colonizing the estuary and those that appear but fail to survive their initial occurrence. Some southern species such as spotfin butterflyfish are fairly consistent as small juveniles in the late summer, when water temperatures are warmest, but they do not survive cool winter temperatures. The clingfish, another new species in the waters around Seven Bridges Road, does not appear in regular time-series collections. The juveniles and adults of this species are captured when they attach themselves to a smooth portion of long-submerged sampling equipment.

Evidence for Sea-Level Rise and its Effects

The effects of a warming world manifest in several ways. First, as the air temperature increases from the accumulation of greenhouse gases in the earth's atmosphere, it causes the accumulated glacier ice to melt, particularly those in the Arctic and Antarctic, thereby releasing more waters

Figure 10.9 "Artoon" of New Seasons at the "Flood Shore" indicating the recognition of increased flooding. By Leslie Ganns from *The Sandpaper*.

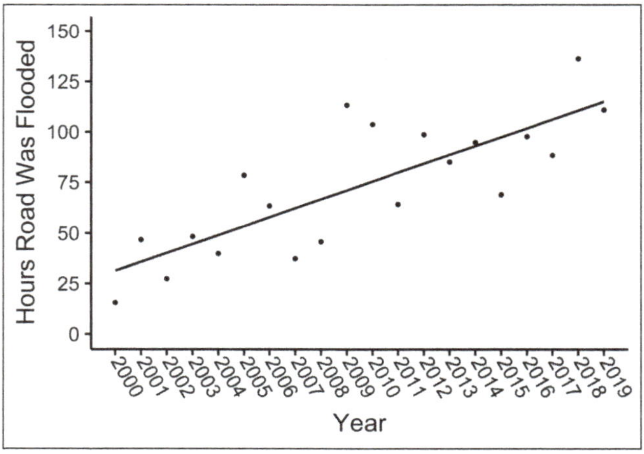

Figure 10.10 Increase in flooding frequency along Seven Bridges Road based on Rutgers University Marine Field Station boat basin tide gauge estimate of number of hours that the road was annually flooded from 2000–2019.

into the oceans and estuaries. The evidence for this is increasing annually. Second, as the ocean water temperature increases, it causes a resultant expansion and increase in overall water volume. Both of these phenomena cause sea-level rise, and are experienced worldwide, including in New Jersey. Even cartoons in a local newspaper frequently recognize the short-term effects, often referred to as "nuisance" or "sunny day flooding" (Fig. 10.9). The Seven Bridges Road area reflects this condition with an increasing frequency in flooding between 2000 and 2019 (Fig. 10.10) and this pattern has continued up to the present. Evidence for this pattern can be found on the frequently changing work schedules at the Rutgers University Marine Field Station due to limited access along the flooded road. Third, the ancient weighty glaciers in the northeastern US, including northern New Jersey, caused the land in much of southern New Jersey to bulge up. Since these glaciers have melted over the last 20,000 years, however, in much of South

Jersey this land is subsiding, called isostatic rebound. Localized effects also contribute to sea-level rise: additional subsidence can occur due to groundwater removal. This is a known phenomenon with the Kirkwood-Cohansey Aquifer due to human population growth and increased water withdrawal in the Atlantic City vicinity.

Extensive studies of the Tuckerton Mound, a shellfish midden that occurs along Seven Bridges Road just west of First Bridge, provides some of the oldest evidence for sea-level rise on the Sheepshead Meadows peninsula (Fig. 2.5). Detailed mapping depicting present-day vegetation topping the midden relative to that below the midden, indicates that the midden was originally much larger, perhaps covering approximating five acres (Fig. 10.11). The mound's lowest level dates to about 2000 years ago. Thus, as sea level has risen, the mound's lower parts have a salt marsh covering with the associated sediments. Today, the exposed portion of the midden measures only about 50 by 100 feet.

Formation of Ghost Forests

Atlantic white cedar, a common tree in the Mullica River watershed, grows only in freshwater bogs or along waterways not exposed to salty waters (Fig. 10.12). As saltwater enters these bogs, either intermittently (as the result of storms that occur with increasing frequency with climate change) or slowly (as sea-level rises, or perhaps due to changes in groundwater levels), these cedars die off, but refuse to rot, and they remain standing for long periods because they are so resilient. These indicator species or "ghost forests" are evident in the Grassle Pocket Marsh along Seven Bridges Road (Fig. 10.13), where the invasive common reed surround them, a pattern occurring in other New Jersey estuaries. Today, the Grassle Pocket Marsh is often exposed to salty waters with high tides, or continuously for days at a time during droughts, or during the summer. Records from 2022 indicate the salinity in Tuckers Creek can range from 0–21 parts per thousand. The cedar trees still standing there consist of Atlantic white cedar and eastern red cedar; the latter is resistant to exposure to low salinity water or salt spray. While relatively easy to tell apart when they are alive, the two species are difficult to identify after they have been dead for some time, i.e. lack leaves, bark, and small branches. Small diameter cores from the trunks serve to differentiate them because the heartwood of eastern red cedar is purple to red while Atlantic white cedar lack coloration (Fig. 10.14). Many of these trees in the marsh itself are small diameter, suggesting that they died fairly young. This exposure

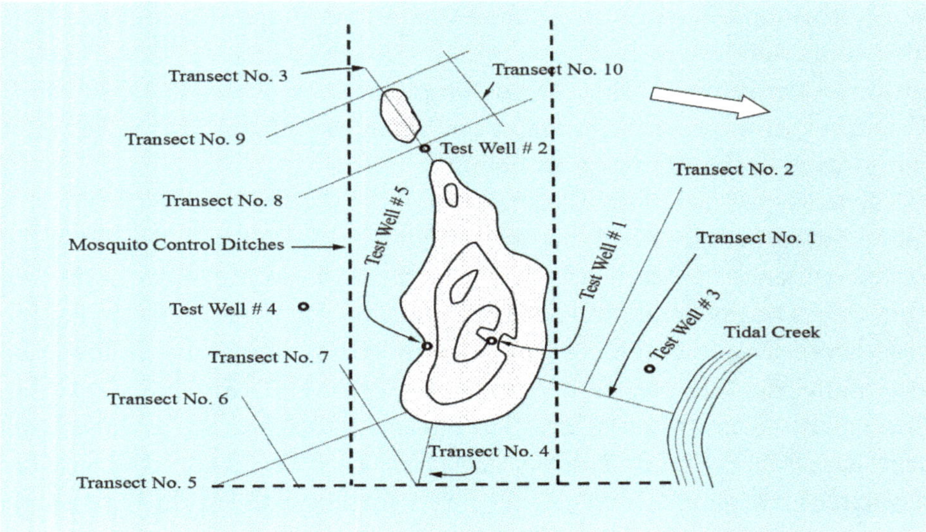

sample transects

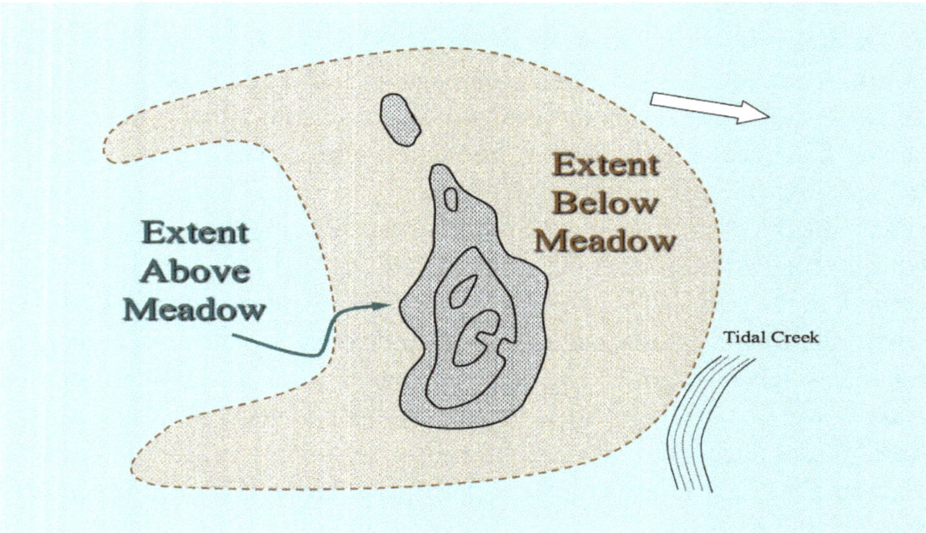

approx. size

Figure 10.11 Diagram of the location and elevation of the Tuckerton Mound (white) relative to linear mosquito ditches (top) with details of the extent above and below the meadow/marsh (bottom). Today the surface of the mound is covered with vegetation similar to that of the adjacent roadside including red cedar trees, poison ivy, and Virginia creeper. A cross section of one of dead trees revealed an age of approximately 200 years thus predating the road by nearly 100 years.

Figure 10.12–14 (Top left) Grove of living Atlantic white cedar trees in upper portion of Grassle Pocket Marsh. (Top right) Adjacent live eastern red cedar and dead Atlantic white cedar in Grassle Pocket Marsh reflecting their relative response to flooding by salty water. (Bottom) Representative differences in cores of Atlantic white cedar (top) and eastern red cedar (bottom) from the Grassle Pocket Marsh.

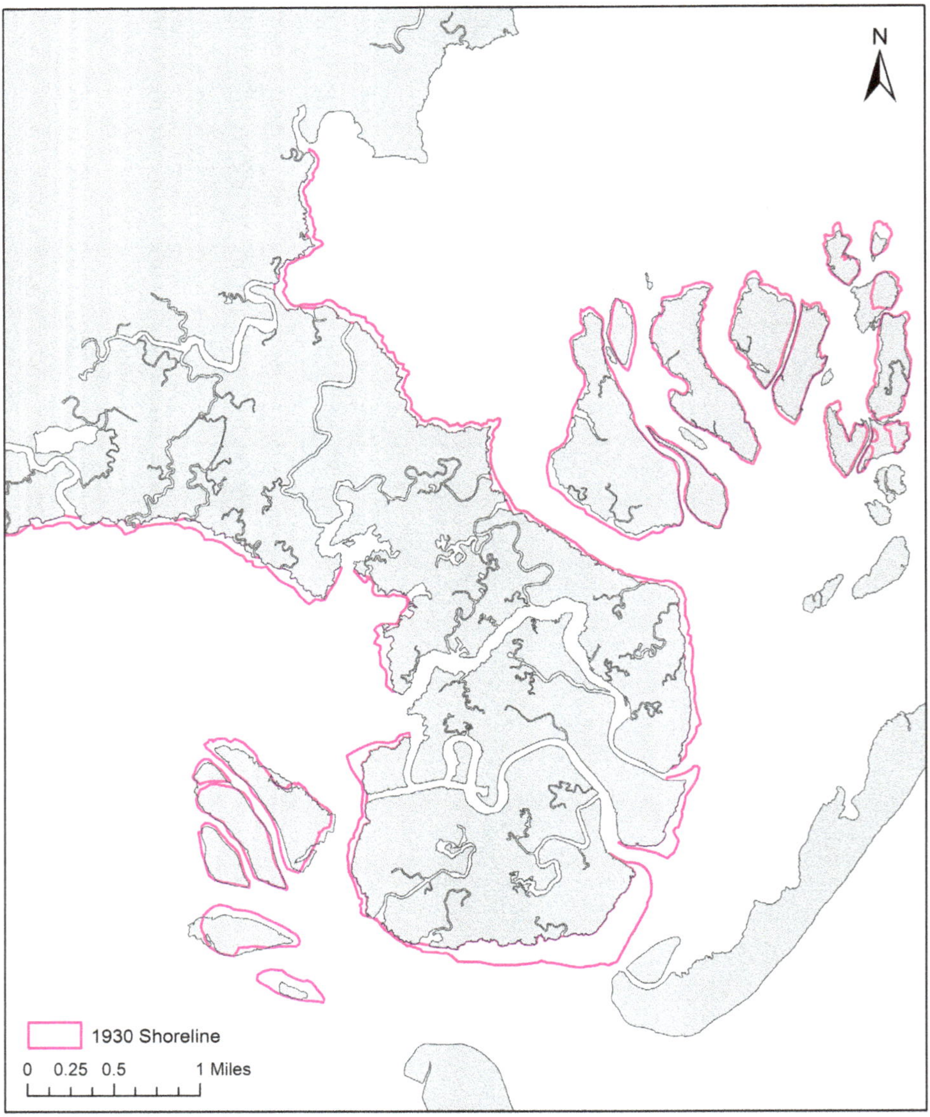

1930 shoreline digitized from 1930 aerial imagery. There are approximately 0.91 square miles of land less in 2022 from 1930.

Figure 10.15 Shoreline change around the Sheepshead Meadows peninsula and the adjacent islands based on aerial imagery from 1930 (red line) to 2015. Note the greatest loss occurred around the peninsula (at the southern edge near Little Egg Inlet) and for selected islands. Prepared by the author and Ryan Larum.

to salty waters determines why all the live trees in the marsh are red cedar, while those that are dead are Atlantic white cedar. This is especially evident when the trunks of both cedars are within inches of each other, but while the white cedars are always dead, the red cedars are always alive (Fig. 10.13). The slightly higher elevations around the marsh's upper edge still host live Atlantic white cedar of considerable age, given their diameter.

In the upper end of nearby Holly Lake, both Atlantic white cedar and eastern red cedar trees are often dead, based on recent observations, which suggests they occurred there naturally when it was a marsh (Chapter 7) before a dam was constructed that continuously flooded the former marsh. The Tuckerton Creek watershed likely hosted other great stands of Atlantic white cedar at an earlier time with a lower sea level. The existence of larger, live Atlantic white cedar in the freshwaters above the dam that forms Lake Pohatcong, in the upper reaches of the Tuckerton Creek watershed (Fig. 5.21) support this interpretation. Live Atlantic white cedar are also common along the Tuckerton Seaport shorelines in an area surrounding the Seaport walkway. These may be survivors due to the freshwater flowing over the dam just upstream.

Marsh Loss and Its Effects on the Sheepshead Meadows and the Adjacent Islands

Based on historical and present-day images and groundtruthing, the marsh has endured extensive loss at the edge and the interior. Multiple evidentiary sources indicate that Sheepshead Meadows are suffering erosion around its edges with a resulting reduction in the marsh's size and therefore production. Analyzing the marsh's areal difference by comparing aerial photos dated 1930 and 2015 indicates extensive erosion and marsh loss around the periphery with the most extensive loss from the mouth of Big Sheepshead Creek to around the southern edge to the west at Jimmies Creek (Fig. 10.15). The loss area includes Hatfield Creek (Fig. 2.6), the site of an early hotel, now underwater (Chapter 2). The same applies to the Willit Hotel (Fig. 2.5), once located at Sheepshead Meadows' southern end. In some instances, it appears intertidal flats replaced the lost marsh, similar to other marshes along the East Coast. A more detailed examination of aerial photos depicting the marsh edge at the deep waters in Shooting Thorofare indicates a loss of 62 linear yards between 1995 and 2015, with much of this occurring by ongoing, typical calving-off of the marsh edge (Fig. 10.16). This is most evident in a close-up view of the area around the

Station between 1930 and 2019 (Fig. 7.2). The same loss at the marsh's edge has also occurred around the adjacent islands. The greatest loss among the Seven Islands includes Fish Island (23.1%), and Tow Island (85.9%). The loss is generally less at the Story Islands complex, with the greatest at West Sedge (22.2%) and East Sedge (62.7%). These losses in the Sheepshead Meadows are consistent with those in many areas along the Northeast Coast of the United States.

The marsh interior is also experiencing loss, possibly due to an inadequate sediment supply to keep up with sea-level rise. The appearance of "rotten spots" in the marsh provides the loss evidence with subsequent substrate breakup leading to shallow, coalescing pools or ponds forming over time. In general, the suspended sediments net import appears insufficient to permit the marsh to gain enough elevation to remain above the sea-level rise. A study detailing suspended sediments flux between Willit Creek, next to the Rutgers University Marine Field Station, and the adjacent waters supplies the evidence along Seven Bridges Road. Other marsh loss stems from an increasing number and size of marsh surface ponds

Figure 10.16 Calving off of marsh edge along marsh thorofare, a common occurrence along Seven Bridges Road.

across Sheepshead Meadows and the adjacent islands, where Open Marsh Water Management or mosquito ditches have not modified the ponds. The sediment that marshes capture comes primarily from the creeks, delivered on the high tides, but the marsh edge can also receive it, especially during spring and storm tides. As the sediment-laden water overflows the marsh, the marsh grasses slow the currents and allow the sediments to drop out and accumulate. As a result, the marsh areas farthest from the creeks and the marsh edge experience flooding less frequently and therefore, typically, receive less sediment. Thus, the marsh cannot keep pace with sea-level rise. Aerial images depicting the meadow provide evidence for this pattern and reveal the formation of larger open water or "lakes" areas (Fig. 10. 17, 10.18, 10.19). The adjacent Seven Islands exhibit this same "lake" formation pattern (Fig. 10.20) and on the Story Island complex (Fig. 7.15, 7.16). Microtidal (<6 feet) marshes most commonly evince this general pattern of salt marsh "runaway fragmentation" in numerous other marshes along the East Coast (Fig. 10.21).

During the fragmentation's initial stages, continuous flooding drowns the saltmarsh cordgrass, causing it to die. These unvegetated areas later deepen as the peats rot that form the marsh's foundation. Again, aerial photographs tell the story showing where larger, deeper ponds have increased fetch that

Figure 10.17 Evidence for the Increase in the size and number of marsh surface ponds until they resemble small "lakes" in the Sheepshead Meadows as seen near the Crab Island fish factory at Seven Islands (top) and Little Sheepshead Creek (upper right). Drone image by Rob Auermuller.

Figure 10.18–19 (Top) Large "lake" along Seven Bridges Road formed between the road (bottom) and Newmans Thorofare consisting of large open ponds and surrounded by shallow pools formed from the flooded marsh. Beyond these, around the periphery, are numerous small discrete ponds and smaller "jigsaw" puddles/ponds that are likely to join the "lake" and increase the size of the larger lake over time and with continued sea level rise. Drone image by Rob Auermuller. (Bottom) Several "lakes" forming on both sides of Seven Bridges Road as the result of sea level rise over time. The location is evident from the abandoned marina (formerly Rand's Marina) and the sandy patch adjacent to the road that is now the "Turtle Garden." The former area of the marsh that is now a "lake" (below the road) can be determined because small remnant islands of former marsh still exist in the center of the lake and especially on the right perimeter. Drone image by Rob Auermuller.

erodes the ponds and "lake" edges. If the ponds become connected with adjacent tidal waters, sediments can reenter the marsh, and marsh vegetation could redevelop, reversing the decay process, provided the marsh can gain enough elevation to keep pace with sea-level rise. Without a steady sediment supply from intertidal and subtidal creeks, and continued tidal overflooding, the "lakes" grow ever larger as they spread across the Sheepshead Meadows peninsula. Other studies suggest that reduced organic material deposits result from reduced established marsh plant abundance and growth. In some instances, small "jig-saw puzzle pools" surround the forming "lakes." These smaller pools will eventually add to the "lake" size (Fig. 10.18, 10.19, 10.21, 10.22). All of these processes result in less marsh grass and thus reduce productivity. At the same time, Open Marsh Water Management increases the overall open water area occupying the marsh surface. The sediments derived from such pond construction are deposited back on the marsh surface (Fig. 7.27).

While relatively infrequent, major New Jersey coastal storms can inflict devastating damage, whether they are subtropical storms, hurricanes, or nor'easters. Numerous accounts exist of lives lost and extreme property damage along the New Jersey coast. Less accumulated knowledge exists,

Figure 10.20 The formation of coalescing ponds or "lakes" is also occurring on Seven Islands. Note the location of the Crab Island fish factory across from Little Sheepshead Creek and Seven Bridges Road (upper right).

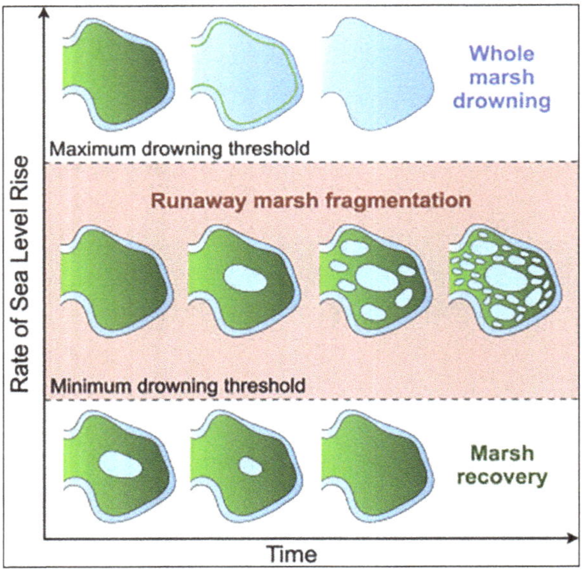

however, concerning these storms' occurrences and impacts to the unoccupied coastal zones, such as near Seven Bridges Road. An exception to the scant information can be found within the sedimentary record that "washovers" leave behind. As storm waters cover the marshes, they deposit overlaying sediments. Brigantine Island's deep cores in the adjacent marsh provide a historical

Figures 10.21–22 (Top) Diagram of "runaway fragmentation of salt marshes" (middle) relative to the entire marsh drowning (top) and full marsh recovery (bottom). From Vinent et al. (2021). (Bottom) Aerial image of a portion of the Sheepshead Meadows crossed by power lines with extensive "jigsaw ponds" being formed on the right of the image with created Open Marsh Water Management ponds and connecting ditches on the left side of the image.

record of past storms that transpired during 1938, 1944, 1950 and 1963, and with some occurring as early as 1821. A photographic record from 1995–2015 exists for the washover immediately adjacent to the Rutgers University Marine Field Station. The images demonstrate, on a smaller scale, the timing and extent that sandy sediments cover the marsh platform until the sediment completely buries the grasses (Fig. 7.2, 7.3). Often these washovers result from major storms like hurricanes, but it is possible that sea-level rise might contribute to increased depositions. A phenomenon known as "sunny day flooding" along Seven Bridges Road serves as a tell-tale reminder of sea-level rise as the flooding does not relate to storms or even high tides (Fig. 10.23).

Influences on the Fauna

Predicting the influences associated with sea-level rise on the many faunal groups that inhabit Sheepshead Meadows might prove difficult, but some insights are becoming available. Studies of insect and spider communities along Seven Bridges Road, along with a general review from other areas, suggest that the dominant marsh grass transformations will influence the community response as the result of changes in sea level, habitat loss, and weather events such as hurricanes. It is difficult, however, to isolate the individual effects within these multiple disturbances.

Figure 10.23 "Sunny day flooding" along Seven Bridges Road. Note Tuckerton in upper center at upland edge. Drone image by Rob Auermuller.

Salt marsh vegetation decline cascades its effects on the marsh surface fauna, based on extensive studies involving the Sheepshead Meadows. One effect of sea-level rise, although less obvious, is a change in the distribution and patchiness of the two dominant plant types along Seven Bridges Road marshes. The low marsh dominant is saltmarsh cordgrass, while in the less frequently flooded high marsh salt hay is dominant. As sea levels rise, saltmarsh cordgrass, with a higher tolerance of frequent tidal flooding, is supplanting the high-marsh salt hay, which cannot survive sustained tidal flooding. These flora changes can directly and indirectly affect the fauna. For example, when salt hay is lost the specialist insects which use this vegetation type are also lost. Further, when salt hay becomes fragmented due to more flooding from sea level rise, insect abundance and composition are altered as salt hay patch sizes reduce and degrade.

The diamondback terrapin, a conspicuous faunal component in Sheepshead Meadows, will likely experience an ongoing habitat reduction with increasing sea-level rise. The elevated habitat along the meadow margins and along Seven Bridges Road, heretofore relatively sandy and dry, serves as nesting sites and nurseries for hatchlings. Visual observations provide evidence these margins now suffer from repeated flooding. The terrapins bury their eggs during mid-summer, and some remain buried through the winter. The nests are susceptible to the standing water and result in many new hatchlings dying. It is likely that optimal nesting sites availability will diminish over time, as has already occurred in other areas like Chesapeake Bay and other coastal zones.

With sea-level-rise-induced marsh overflooding, the overall marsh surface and the habitat it provides will experience shrinkage, limiting the nesting area and the wrack the hatchling terrapins use. An observable exception is the frequent washovers that have converted the marsh into supratidal sandy habitats and might increase nesting habitat in specific areas along the marsh edge including in a small area adjacent to the Rutgers University Marine Field Station. As an example, in one day, July 4, 2022, I observed three female terrapins coming ashore on the washover and digging nests during one high tide cycle.

In summary, the salt marshes on the Sheepshead Meadows provide numerous ecosystem services, including habitat and food web support, buffering against storm wave damage, shoreline stabilization, flood water storage, enhancing water quality, preserving biodiversity, carbon storage and socio-economic services including ecotourism, aesthetics, and education.

The effects from climate change and sea-level rise can influence all these ecosystem services.

Speculations and Questions about the Future

We can often gain insights into the future through examining the past, and particularly as it applies to the area around Seven Bridges Road. Referencing historical baselines becomes especially critical, allowing us to interpret change and thus avoid the influence of shifting baselines. That is, when succeeding generations of observers subconsciously recall their youthful environment as "natural" and ignore prior changes (sometimes referred to as "generational amnesia"). Rapid climatic changes and sea-level rise will likely change that viewpoint as changes become obvious to the most careful observers. It is critical, however, to understand the how and why behind these changes now occurring.

During the summer of 2023, southern New Jersey and the world experienced the warmest air temperatures in recorded history. Prior decades of sampling demonstrate that rising air temperatures will cause increased water temperatures, affecting the larval and eventually the adult fish fauna due to the more frequent influx of southern species and the decline of northern species in Sheepshead Meadows plankton samples. These southern species could include rough silverside, southern kingfish, pink wormfish, and striped burrfish, while some northern species such as American lobster will decline. Some commercially important southern shrimp species, such as brown and white shrimp, have already begun appearing in our bottom trawl samples around the Sheepshead Meadows during the past decade.

Some locations elsewhere in New Jersey, and along the East Coast of the United States, will experience marshes migrating and invading adjacent uplands due to changing climate and sea-level rise. This cannot occur in the Sheepshead Meadows because the mainland edge of the peninsula features cultural adaptations, comprising lagoon and upland developments, including a golf course (Fig. 10.24). Thus, "coastal squeeze" has already occurred on the edge of the peninsula. Sea-level rise may be the cause of reductions of the number of islands (from seven to five) at "Seven" Islands and the loss of area in these islands and those in the Story Island complex. The likelihood that these effects of sea-level rise are going to continue to happen is supported by the occurrence of ghost forests of Atlantic white cedar in the pocket marshes and elsewhere in the Mullica Valley.

The increasing marsh pool numbers, extent, and "lakes" on the marsh platform provide anecdotal sea-level rise evidence, now obvious along Seven Bridges Road. The extensive "lakes" are most visible on the road's south side and on both sides of Little Thorofare. The practice of Open Marsh Water Management, marsh alteration to control mosquito populations, curiously also creates large areas of open water (Fig. 10.22). Do these mosquito control efforts also reduce marsh plant production in the same way that sea-level rise does as it drowns the marsh surface? This is a question that deserves more attention, especially as sea-level rise continues increasing.

Further changes will likely occur into the future. Sea-level rise, combined with continuous subsidence, suggests ongoing marsh edge erosion, that tidal creeks will become wider and longer, and that marsh surface "lakes" will increase in number and size, resulting in declining marsh size. These changes will occur as sediments fail to keep up with sea-level rise while subsidence continues. All of these will increase marsh fragmentation and decrease marsh area and productivity. There may be exceptions to these patterns. The thin vegetation edges along marsh creeks may continue to collect sediments, due to more frequent flooding, and may still be present. Storms could continue to provide silty sediments and even sand carried within some marsh edges washovers, similar to that occurring on the edge of Shooting Thorofare at the Rutgers University Marine Field Station. In fact, some recent studies involving East Coast marshes suggest that marshes with lower tidal ranges (less than 6 feet change), such as in the Sheepshead Meadows area (about 3 feet change), will continue shrinking or disappear completely with sea-level rise because the sediment supply, either from inorganic or organic (breakdown of marsh grasses), will be insufficient for sustaining marsh stability. A FloodMapper projection supports this interpretation, suggesting that all the marshes on the Sheepshead Meadows and adjacent islands will experience flooding on every high tide (mean high, high water) with just a one-foot rise in sea level (Fig. 10.24). Sea-level rise modeling and the resulting marsh loss along the New Jersey coast also suggests a concern for marsh nesting birds. Estimates indicate that only 10–15% of tidal marshes will survive, and thus several species (clapper rail, Nelson's sharp-tailed sparrow, saltmarsh sharp-tailed sparrow, seaside sparrow, willet) will lose their marsh habitat (>85%) by 2050.

This frequent flooding, especially at lower elevations (prior Fig. 9.7), may cause erosion of the road. An exception would be areas where the road is at a slightly higher elevation, such as north of First Bridge and south of Little Sheepshead Creek. Other exceptions include the fish factory area on Crab Island, the inclined approaches at each bridge, and the Story Island, and Tuckerton Cove dredge spoil disposal sites. The projected sea-level rise and flooding (Fig. 10.24) will likely cause the marsh platform vegetation to disappear and be replaced with unvegetated tidal flats throughout the Great Bay Wildlife Management Area. Studies in adjacent Delaware Bay

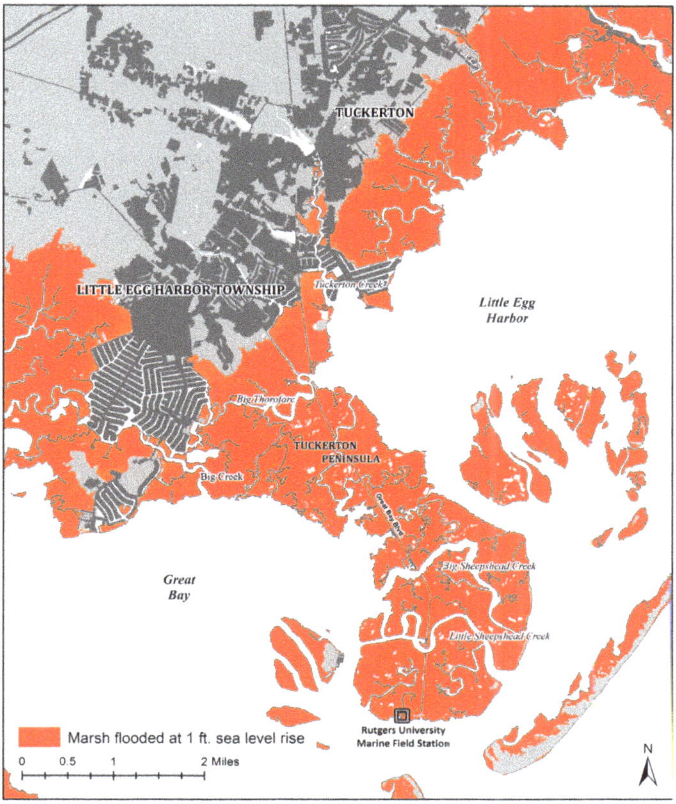

MHHW flooding on Tuckerton Peninsula, NJ given one foot of sea level rise. Sea level data from NOAA Coastal Services Sea Level Center. Dark Grey areas are urban areas from NJDEP land use data.

Figure 10.24 Effect of anticipated one foot of sea level rise on Sheepshead Meadows and adjacent islands based on New Jersey Department of Environmental Protection FloodMapper. Dark grey areas are urban areas in Little Egg Harbor Township and Tuckerton based on New Jersey Department of Environmental Protection land use data. Prepared by Ryan Larum.

and Barnegat Bay calculate that the time to total marsh submergence is 60–80 years, with some predicted to submerge in as few as five years. Some marsh areas are already submerged at the marsh edge with the formation of "lakes." Further, the intertidal marsh platform may lose the depressions and the pools that serve as nurseries and winter refugia for a variety of killifish, including the ecologically important mummichog, and the multiple species of birds which often focus their feeding on the pools.

In the more distant future, given the forecasted marsh drowning with increasing sea-level rise, will the marsh platform evolve into intertidal flats, as has already occurred at some locations (Fig. 10.25)? Such change can currently be witnessed on portions of the Little Sheepshead Creek margins, at the mouth of the Rutgers University Marine Field Station boat basin, and at selected islands in the Story Island complex and at Crab Island (Fig. 10.15). These intertidal sites typically comprise muddy sand or fully sand, depending on location. The lower elevation intertidal and subtidal creeks channels will probably remain visible on very low tides,

Figure 10.25 Intertidal flats with sea lettuce macroalgae on the surface adjacent to patches of saltmarsh cordgrass in Little Sheepshead Creek (left). With further sea level rise the cordgrass will be replaced by more of the intertidal flat. Same habitat type adjacent to the Rutgers University Marine Field Station (right). This area was salt marsh in the 1930s based on aerial photography.

but it is undeterminable how the marsh-dependent fish, crabs, shrimp, and mussels will respond to these altered habitats. One scenario suggests that the tidal flats, especially those colonized by macroalgae (sea lettuce) (Fig. 10.25), will serve some of these functions, as they do for blue crabs (Chapter 8), but to some diminished degree. They may not be, however, as significant a source for detritus as the extensive, three-dimensional vegetated marshes.

Frequent or nearly continuous current marsh surface flooding presents the possibility of actually reducing mosquito egg laying and larval survival if the marsh fails to migrate, as in the Sheepshead Meadows case. Predictive modeling from other areas certainly suggests that the transition from salt marshes to intertidal flats will occur. It is doubtful, however, that the flats will be as ecologically productive as salt marshes, especially in their value as nurseries for shrimp, crabs, terrapins, and fish.

An unresolved potential result from climate change is hypoxia or lower-than-normal oxygen levels. Such a condition often results with anthropogenic changes, including higher temperatures and eutrophication. The deeper waters of Chesapeake Bay exhibit these problems, but the shallow waters in and around the Sheepshead Meadows lack such evidence. The exception can be found within the marsh surface pools, even natural ones, which retain very high oxygen levels due to photosynthesis during the day, but can drop to levels near zero (anoxia) during the night, and especially near morning. The simultaneous effect of acidification, due to increasing carbon dioxide in the atmosphere, can lower the pH of estuarine waters, but also increases fish sensitivity to hypoxia, especially for species of silversides abundantly found in the Sheepshead Meadows area. A common response to low dissolved oxygen is water surface skimming, which makes these fish more available to bird predation, as frequently happens during the summer.

Sea-level rise will change some features along the road. The common reed and some shrubs and trees will likely die and disappear along the roadside, with more frequent and higher saltwater flooding. Slightly higher elevations, including the dredge spoil disposal facilities (CDFs) adjacent to Tuckerton Creek (Fig. 7.32, 7.33, 7.34), will eventually become "islands." High salinity intolerant flora such as common reed, shrubs, and trees will only flourish on the berms' higher portions, and on the higher elevations of bridge approaches (Fig. 7.20).

Baselines for Future Evaluations

As indicators of climate change and sea-level rise effects, the value that Sheepshead Meadows and the adjacent islands offer are numerous and ongoing. In part, this is due to much of the area's naturalness, thus providing an excellent baseline for evaluating change due to climate change and sea-level rise, and as a reference for restoring habitats and mitigation efforts. These "ecology of place" studies further enhance this work, particularly with the Rutgers University Marine Field Station and the Jacques Cousteau National Estuarine Research Reserve being located within the area of interest.

Current indicators of climate change and sea-level rise include:

1) evidence for ghost forests of Atlantic white cedar,
2) marsh loss at the edge (in the meadows and adjacent islands),
3) marsh loss in the interior of the meadows and the islands (formation of "lakes" with aerial imagery from helicopter and drones),
4) shifting margins for saltmarsh cordgrass and salt hay,
5) frequency of road flooding, and
6) change in the fish and insect fauna.

Other advantages the Sheepshead Meadows offer include marsh restoration and mitigation efforts that provide insights into marsh responses. Numerous attempts to restore or mitigate salt marsh loss due to sea-level rise have occurred at other sites with thin-layer sediment placement. In the Sheepshead Meadows, we have examined the beneficial use of dredged sediments. A prior study, based on a long-term fish-trapping study (Table 10.1) in the boat basin before, during, and after the dredging found little effect in the kinds and number of fish. The nearby marsh surface received some sediments from a later dredging (Fig. 10.26, top). The study emphasized fiddler crabs in evaluating the response because they are critical marshes components. These crabs act as bioengineers during their burrowing (Fig. 10.26, top right) at all life-history stages, and serve as important salt marsh food web links. Observations during the evaluation found that the mud fiddler quickly colonized the dredged sediment deployed on the marsh surface. In general, the marsh vegetation responded quickly and continuously over the three-year observation period (Fig. 10.26, bottom). Thus, this metric demonstrates that adding dredged sediments provides one way in which salt marsh vegetation becomes more resilient and offers local fauna a habitat.

Figure 10.26 Image of Thin Layer Deposition (TLD) experimental sites created adjacent to the Rutgers University Marine Field Station from dredged sediments (top left), as colonized by fiddler crab burrows (top right), and eventually saltmarsh cordgrass within one year later (bottom).

The Sheepshead Meadows baseline values provide a study model for other sites in evaluating future climate change and sea-level rise effects. Several ongoing metrics further enhance this study model. These enhancements include prior and ongoing studies of:

1) sea-level rise in the region, some of the world's fastest due to sea-level rise in combination with subsidence,
2) LIDAR based mapping of the Sheepshead Meadows elevation,
3) frequent and extensive flora and fauna treatment in these natural and partially altered habitats, including extensive fish fauna synthesis, and

4) some of the longest-time series on the East Coast based on plankton sampling for the early life history stages of fish and crabs; trap sampling for juveniles and adults; and otter trawl sampling for juveniles and adults.

For the marshes, these include:

1) monitoring surface elevation tables,
2) vegetation transects,
3) ground water wells,
4) digital elevation models on the same vertical datum, all at the Sheepshead Meadows Sentinel Site by the Jacques Cousteau National Estuarine Research Reserve,
5) the prior experimental evaluations concerning the influence of flooding and draining on marshes,
6) evaluation of Thin Layer Deposition as a means of enhancing marsh resilience, and
7) continuing extensive and intensive studies of the basic insects and spiders ecology over decades.

Together, the above work provides the ideal location for studying "conservation behavior": that is, realizing the importance that animal behavior plays in understanding and mitigating threats to wild populations, given the threats facing marsh systems such as the Sheepshead Meadows.

For the waters, these include:

1) system Wide Monitoring Program (SWMP) data loggers by the Jacques Cousteau National Estuarine Research Reserve,
2) numerous other water quality measurements (temperature, pH, dissolved oxygen, turbidity) associated with fish and crab sampling, and
3) relevant broad scale measures (water temperature, pH, etc.) in adjacent coastal waters with remotely operated vehicles.

Table 10.1 Changes in fish and crab abundance (increasing, declining, no change; blank space = insufficient data) from different time series sampling programs in waters in and around the Sheepshead Meadows. Sampling for larvae (plankton net weekly, year round), juveniles and adults with traps (2x per week, year round), juveniles and adults with trawl (July and September) were the basis for the evaluation. Source indicates Northern = N, Southern = S, Local = L, Invasive = Inv.

	Source	Larvae	Juveniles and Adults	
Common Name	(L,N,S,Inv)		Traps	Trawl
American eel	S	No change	No change	Declining
Asian Shore crab	Inv		Variable	
Atlantic menhaden	L	Increasing		
Atlantic croaker	S	Increasing		No change
Atlantic herring	N	Declining		
Atlantic silverside	L	Increasing	No change	
Atlantic thread herring	S	Increasing		
Black drum	S			
Black sea bass	L	No change	No change	No change
Blackcheek tonguefish	S	Increasing		
Blue crab	L	No change	No change	
Conger eel	S	No change	No change	
Cunner	N	No change	Declining	No change
Darter goby	S	Increasing		
Emerald sleeper	S	Increasing		
False silverstripe halfbeak	S	Increasing		
Fat sleeper	S	Increasing		
Four spine stickleback	L		No change	Declining
Gag	S		No change	
Green crab	Inv		Declining	
Green goby	S	Increasing		No change
Grey snapper	S			
Grubby	N		Declining	Declining
Highfin goby	S	Increasing		
Ladyfish	S	Increasing		
Mummichog	L		No change	
Naked goby	L	No change	No change	No change

Northern pipefish	L	No change	No change	Declining
Oyster toadfish	L	No change	Declining	No change
Pinfish	S	Increasing		
Pollock	N	Declining	Declining	
Rock crab	N		Declining	
Sand lance	N	Declining		
Seaboard goby	L	Increasing	Increasing	
Sheepshead minnow	L		No change	
Silver anchovy	S	Declining		
Silver perch	S	Increasing	Increasing	Increasing
Smallmouth flounder	L	Declining		No change
Spot	S	Declining		
Spotfin butterflyfish	S		No change	
Spotted hake	L		Declining	Increasing
Striped anchovy	L	Increasing		No change
Striped blenny	S	Increasing		
Striped killifish	L	Declining	Declining	
Summer flounder	L	Increasing		
Tarpon	S	Increasing		
Tautog	L	Declining	Declining	No change
Three spined stickleback	N	Declining	Declining	
Weakfish	S			Declining
Windowpane	L			Increasing
Winter flounder	N	Declining	Declining	No change

Appendix Table 1. Common and scientific names of flora and fauna included in the main text

Common Name	Scientific Name
Flora	
Arrowwood	Viburnum dentatum
Agardh's red weed	Agardhiella sublata
Atlantic white cedar	Chamaecyparis thyoides
Bayberry	Myrica pennsylvanica
Black cherry	Prunus perseca
Common reed	Phragmites australis
Dodder	Cuscuta compacta
Eelgrass	Zostera marina
Eastern red cedar	Juniperus virginiana
Glassworts	Salicornia spp.
Graceful red weed	Gracilaria tikvahiae
Green Fleece	Codium fragile
Greenbrier	Smilax rotundifolia
Groundsell	Brecharis halimifolia
Hollow green weed	Ulva intestinalis
Marshelder	Iva frutescens
Marsh mallow	Hibiscus palustris
Poison ivy	Rhus toxicodendros
Queen Anne's Lace	Daucus corata
Red cedar	Juniperus virginica
Red fescue	Festuca rubra
Saltmarsh cordgrass	Spartina alterniflora*
Salt hay	Spartina patens*
Salt grass	Distichlas spicata
Sassafras	Sassafras albidum
Seaside goldenrod	Solidago spp

Sea lavender	Limonium nashi
Sea lettuce	Ulva lactuca
Switch-grass	Panicum virgatum
Virginia creeper	Parthenocissus vitacea
Widgeon grass	Ruppia maritima
Winged sumac	Rhus copallinum

Fauna

Chelicerates

Horsehoe crab	Limulus polyphemus

Crabs and Shrimp

American lobster	Homarus americanus
Blue crab	Callinectes sapidus
Brackish grass shrimp	Palemonetes mundusnouvus
Brown shrimp	Farfantepenaeus azteus
Daggerblade shrimp	Palaemon pugio
False zostera shrimp	Hippolyte pleuracanthus
Ghost shrimp	Gilvossius setimanus
Green crab	Carcinus maenas
Japanese shore crab	Hemigrapsis sanguineus
Lady crab	Ovalipes ocellatus
Oriental shrimp	Palaemon macrodactylus
Marsh crab	Sesarma reticulatum
Mud crabs	Scylla serrata
Marsh grass shrimp	Palaemon vulgaris
Mud fiddler crab	Uca pugnax
Sand fiddler crab	Leptuca pugilator
Sand shrimp	Crangon septumspinosa

Insects

Black salt marsh mosquito	Aedes taeniorhynchus
Eastern salt marsh mosquito	Aedes sollicitans
Greenheads	Tabanus nigrovittatus
Leafhoppers	Family Cicadellidae
Planthoppers	Family Delphaenidae
Monarch butterfly	Danans plexippus
Pandorus sphinx	Enmorpha pandorus

Shellfish

Blue mussels	Mytilus edulis
Oyster	Crassostrea virginica
Hard clam	Mercenaria mercenaria
Sea scallop	Placopecten magellanicus
Slipper shells	Crepidula fornicata
Ribbed mussel	Geukensia demissa

Fish

American eel	Anguilla rostrata
American sand lance	Ammodytes americanus
Atlantic cod	Gadus morhua
Atlantic croaker	Micropogonias undulatus
Atlantic herring	Clupea harengus
Atlantic menhaden	Brevoortia tyrannus
Atlantic needlefish	Strongylura marina
Atlantic silversides	Menidia menidia
Atlantic thread herring	Opisthonema oglinum
Banded killifish	Fundulus diaphanus
Bay anchovy	Anchoa mitchilli

Black drum	Pogonias cromis
Black sea bass	Centropristis striata
Blackcheek tonguefish	Symphurus plagiusa
Bluefish	Pomatomus saltatrix
Bluegill	Lepomis macrochirus
Clingfish	Gobiesox strumosus
Cobia	Rachycentron canadum
Conger eel	Conger oceanicus
Crevalle jack	Caranix hippos
Cunner	Tautogolabrus adspersus
Darter goby	Ctenogobius boleosoma
Eastern mudminnow	Umbra pygmaea
Emerald sleeper	Erotelis smaragdus
False silverstripe halfbeak	Hyprohamphus meeki
Fat sleeper	Dormitator maculatus
Feather blenny	Hypsoblennius hentz
Fourspine stickleback	Apeltes quadracus
Gag	Mycteroperca microlepis
Goatfish	Mullidae
Gray snapper	Lutjanus griseus
Green goby	Microgobius thalassinus
Grubby	Myoxocephalus aenaeus
Hickory shad	Alosa mediocris
Highfin goby	Gobionellus oceanicus
Inland silversides	Menidia beryllina
Inquiline snailfish	Liparis inquilinus
Inshore sand lance	Ammodytes americana
Ladyfish	Elops saurus
Largemouth bass	Micropterus salmoides

Lined seahorse	Hippocampus erectus
Lookdown	Selene vomer
Mummichog	Fundulus heteroclitus
Naked goby	Gobiosoma bosc
Northern kingfish	Menticirrhus saxatilis
Northern pipefish	Syngnathus fuscus
Northern searobin	Prionotus carolinus
Northern sennet	Sphyraena borealis
Oyster toadfish	Opsanus tau
Pinfish	Lagodon rhomboides
Pollock	Pollachius virens
Pumpkinseed	Lepomis gibbosus
Rainwater killifish	Lucania parva
Scup	Stenotomus chrysops
Seaboard goby	Gobiosoma ginsburgi
Sheepshead	Archosargus probatocephalus
Sheepshead minnow	Cyprinodon variegatus
Silver anchovy	Engraulis eurystole
Silver perch	Bairdiella chrysoura
Skilletfish	Gobiesox strumosus
Smallmouth flounder	Etropus microstomus
Smooth dogfish	Mustelus canis
Speckled worm eel	Myrophus punctatus
Spot	Leiostomus xanthurus
Spotfin butterflyfish	Chaetodon ocellatus
Spotfin killifish	Fundulus luciae
Spotted hake	Urophycis regia
Striped anchovy	Anchoa hepsetus
Striped bass	Morone saxatilis

Striped blenny	Chasmodes bosquianus
Striped burrfish	Chilomycterus schoepfi
Striped cusk-eel	Ophidion marginatum
Striped killifish	Fundulus majalis
Striped searobin	Prionotus evolans
Summer flounder	Paralichthys dentatus
Tarpon	Megalops atlanticus
Tautog	Tautoga onitis
Toadfish	Opsanus tau
Threespine stickleback	Gasterosteus aculeatus
Weakfish	Cynoscion regalis
White mullet	Mugil curema
Windowpane flounder	Scophthalmus aquosus
Winter flounder	Pseudopleuronectes americanus

Reptiles

Diamondback terrapin	Malaclemys terrapin
Eastern box turtle	Terrapene carolina carolina
Loggerhead	Caretta caretta
Snapping turtle	Chelydra serpentina
Spotted turtle	Clemmys guttata

Birds

American bittern	Botaurus lentiginosus
American black duck	Anas rubripes
Bald eagle	Haliaeetus leucocephalus
Barn swallow	Hirundo rustica
Belted kingfisher	Megaceryle alcyon
Black bellied plover	Haliaeetus leucocephalus

Black duck	Anas rubripes
Black vulture	Corgyps atratus
Black-crowned night heron	Nycticorax nycticorax
Boat-tailed grackle	Qusicalus major
Brant	Branta bernicla
Canada goose	Branta canadensis
Clapper rail	Rallus longirostris
Common loon	Gavia immer
Curlew	Mumenius americanus
Dowitcher	Limnodromus
Glossy ibis	Plegadis falcinellus
Great blue heron	Ardea herodias
Great egret	Ardea alba
Greater yellowlegs	Tringa melanoleuca
Harrier	Circus cyaneus
Herring gull	Larus argentatu
Kingfisher	Alcedinidae
Laughing gull	Leucophaeus atricilla
Least sandpiper	Calidris minutilla
Little blue heron	Egretta caerulea
Mute swan	Cyngus olor
Nelson's sharp-tailed sparrow	Ammedramus nelsoni
Osprey	Panidon haliaetus
Oystercatcher	Haematopus
Plover	Subfamily Charadriinae
Red knot	Calidris canutus
Red-winged blackbird	Agelaius phoeniceus
Ruddy turnstone	Arenaria interpres
Saltmarsh sharp-tailed sparrow	Ammodramus caudacutus

Sanderling	Calidris alba
Seaside sparrow	Ammodramus maritimus
Snowy egret	Egretta thula
Semipalmated plover	Charadrius semipalmatus
Semipalmated sandpiper	Calidris pusilla
Tree swallow	Tachycineta bicolor
Tricolored heron	Egretta tricolor
Turkey vulture	Cathartes aura
Whimbrel	Numenius phaeopus
Wild turkey	Meleagris gallopavo
Willet	Tringa semipalmata

Mammals

American mink	Neovision vison
Atlantic bottlenose dolphin	Tursiops truncatus
Black bear	Ursus americanus
Gray squirrel	Sciurus carolinensis
Gray wolf	Canis lupus
Groundhog	Marmota monax
Harbor seal	Phoca vitulina
Manatee	Trichechus manatus
Mink	Mustela vison
Opossum	Didelphis Virginiana
Raccoon	Procyon lotor
Red fox	Vulpes vulpes
River otter	Lontra canadensis
White tailed deer	Odocoileus virginianus

Footnotes:

1. Recent reclassification of Spartina indicates different taxonomic views of this genus as follows:

   Paul Peterson, et al., 2014. A molecular phylogeny and new subgeneric classification of *Sporobolus* (Poaceae: Chloridoidae: Sporobolinae). *Taxon, 63*(6), 1212–43.

   Alejandro Bortolus et al. 2019. Supporting *Spartina*: interdisciplinary perspective shows *Spartina* as a distinct solid genus. *Ecology, 100*(11), e02863.

Appendix 1

Coastal Late Woodland Sites in Mystic Island, New Jersey (Osborne Island, Graveling Point, and Wells Islands area)

Andrew Stanzeski and John Stanzeski

Introduction

The following is a short description of the Late Woodland sites found along a portion of the central New Jersey coast. The archaeological work for this paper was conducted in Ocean, Burlington, Atlantic and Cape May counties in New Jersey. A majority of the work was done in the Mystic Island area (Osborne Island, Wells Island and Graveling Point) Ocean County, New Jersey. This area can be referred to as the Mystic Island Complex.

Important sites are being ignored because of an inability to identify prehistoric sites along the coast. In general, all prehistoric sites from Paleo to Contact have the same problem. But the Late Woodland sites are the most numerous, and perhaps the most easily identifiable because of the presence of shell. Other historic and prehistoric time periods found in the area are reviewed. An example from one site indicates the differences in dates when using carbon (wood) and shell to date a site. Also, the possible use of sediments below a marsh to give an approximate date for the bottom sediments is described.

Background

To date there is a lack of information regarding Late Woodland sites along the southern coastal region of New Jersey (Stanzeski, in review). Survey work done earlier by other archaeologists (Skinner and Schrabisch, 1913; Cross ,1941) showed a relatively small number of prehistoric sites along the coast compared to survey work done throughout the rest of New Jersey.

Late Woodland sites found along the central coastal region of New Jersey in the area of Wells Island, Osborne Island and Graveling Point are typical of Late Woodland sites found along the rest of the coast of New Jersey. Between 1972 and 2020 in the area of Mystic Island there were different stages of clearing the area for housing developments and making lagoons for waterways. Several sites were located and recorded. These are identified by Smithsonian Institution and New Jersey numbers such that 28 is for the state, OC is for Ocean County and the last number is the site number in the county.

Identification of the Sites

On Wells Island (Fig. 1) two prehistoric sites were found. The sites were identifiable because of the presence of shell. One of the shell middens on the southwest end of the island (28OC175, Site 1) was adjacent to the meadows on the high ground (Fig. 1) and was made up of a thin scattering of hard clams (*Mercenaria mercenaria*), covering a circular area ten feet by

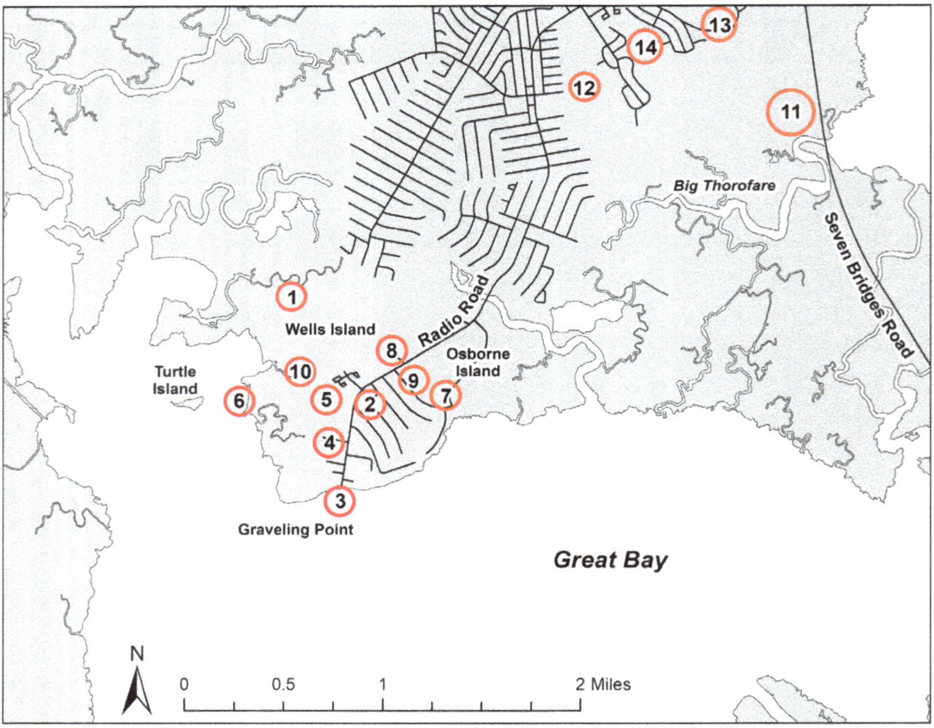

Figure 1 Location of primary study sites (1–10) in the vicinity of Wells and Osborne Islands near the connection between Great Bay and the Mullica River. Other sites include the Tuckerton Mound (Site 11), the Jillson Farm Site (12), the Penella Site (Site 13) and the Willits/Andrews Farmstead (Site 14).

fifteen feet. The shell midden was most likely of Middle or Late Woodland origin and probably a short-term shell fishing episode. One bannerstone fragment was found one hundred feet north of the site (Fig. 2).

In the southeastern part of Wells Island (28OC101, Site 2) another site was located in a lagoon development cut. Oyster shell (*Crassostrea virginica*) recovered from the site were dated to 650+45 B.P., a Late Woodland date (Stanzeski, 1996). Artifacts found at this site included pottery, a jasper triangle, a jasper drill, jasper and chert flake tools, and clam shell tools (Fig. 2). All lithics were of pebble material and potsherds were fabric impressed. This site was located in the meadows where a lagoon was cut out for a housing project, disturbing part of a midden. Soils from the road fill and dredge material covered the midden and the meadows. Below the fill were one to two feet of buried meadow material (Level 1). Level Two, below the meadows, contained the midden material, 1"–13" thick of black sand and shell. The midden was composed mostly of oyster shells, with a few clams and other shells. In Level Three, there was light yellow sand. In this level, east in the lagoon cut and away from the midden, Archaic artifacts were found in the subsoils at low tide (Massler, pers. comm.). The lagoon cut showed that the midden had a width of 50 to 100 feet. No features were located below the shell midden area. This was the location of a gorget and

Figure 2 A piece of pottery and bannerstone retrieved at Wells Island (Site 2).

burials found and reported by Bello and Eisenberg (1988). The human skeleton sample was of six individuals, three adults, two sub adults, and one young adult. The remains were possibly a "one family grave site." The found gorget had three holes and was made of Steatite, overall length was 2 inches (Fig. 3). This site was most likely part of Dr. Cross's site location (Cross, 1941) that is just across the road in the woods above the meadows.

Figure 3 A gorget, 2 inches in length, found in the southeastern point of Wells Island.

Between Wells Island and Graveling Point on Radio Road (Fig. 1) three more sites were located. One location was the Graveling Point Site (28OC176, Site 3) dating from the Archaic, Transitional and Woodland Periods based on artifacts from those periods. Artifacts were found on the beach and in a lagoon cut (Fig. 4). Also, more recent artifacts found were lead weights, ships ballast, ceramics, and glass. In one area yellow brick, posts, and a barrel (used for a well, privy or bait holding tank) were found (Wilson, 2020, pers. comm).This most likely was the location where the British landed in 1778. That led to the massacre of Pulaski's Legion (Kamp, 1966). This site is partly submerged at high tide. After 30 years most of this site has been washed away and the lagoon cut has been silted in. The other site, an oyster shell midden location (28OC94, Site 4), was in-between two lagoon cuts between Graveling Point site and Wells Island. You can only see this midden at low tide in the lagoon cut. The site is covered with dredge material dug up from the lagoon with oyster in the dredge material. South of Wells Island, Archaic and Late Woodland artifacts (Fig. 2) were found (28OC173, Site 5) This site is north of the previous site, in the meadows on a small island in a mosquito ditch cut (Massler, pers. comm.). This would not have been found except for a ditch that was excavated though the island for mosquito control. In walking the ditch on the island, fire cracked rocks were found. The island is one of four small islands that are above the marsh in a crescent pattern. The islands are remnant ridges or dunes that were formed in the past, possibly relic Quaternary deposits

Figure 4 Artifacts recovered from Graveling Point (Site 3). Top row, left to right: two broad spears, Triangular Point, Bart Island Point, Rossville Point, Orient Fishtail Point. Bottom row, left to right: potsherd, pendant, Gorget fragment, soapstone body sherd.

formed at the mouths of rivers. This would be during the early stages of the Mullica River formation.

In the meadows north of Graveling Point on the Mullica River (between Graveling Point and Turtle Island) (Fig. 1) (28OC246, Site 6), artifacts were found in the water and at low tide on the beach. Most of the artifacts are Late Archaic, Bare Island and Poplar Island Points (Massler, pers. comm.). This site, like other early sites, are under water. The shore line during the Early Archaic and Paleo Indian period would be miles from the present shore line. Remains of large animals such as caribou have been found in approximately 50 feet of water 3 miles off Beach Haven inlet (Parris et al., 1989).

In the early 1970s on Osborne Island (Fig. 1) (28OC76, Site 7) another oyster midden was located during the construction of a golf course and housing development. Most of this midden was in the plow zone, little was left in the subsoil. Part of the midden had been bulldozed into a pile for landscaping. No features at the time were located below the plow zone. The oyster midden that was still intact covered an area over two hundred fifty feet around. Artifacts recovered were a broken chert triangle, jasper and chert flakes made from pebble material, fire cracked rock and fabric impressed pottery. This site was part of a larger site that covers the south end of the island. In the center of the island on higher ground, Late Archaic material has been found. On the west end on the island across Radio Road (Fig. 1) another site was located (28OC174, Site 8) with Late Woodland pottery, jasper flakes, and shell when houses were constructed.

On both Wells Island and Osborne Island (Fig. 1) individual jasper and chert flakes have been found hundreds of feet away from the oyster middens. The flakes are small in number, two to five flakes scattered around an area of five to ten feet. This may be evidence of a short-term gathering or station associated with the main site. Wells Island, Osborne Island, and Graveling Point make up a total of many Late Woodland sites that are large and small with the main activity in the Late Woodland for shellfishing.

Folklore and the Historic Period

There were numerous early activities, folklore, and historic houses in the area of Mystic Island. On Wells Island there is a legend of a massacre of the Egg Harbor tribe by the Leeds Point tribe. Both had hostility for one another. In return, friends of the Egg Harbor tribe, the Manahawkin Tribe, hearing of the massacre the next day came to Wells Island and massacred the Leeds Point tribe (Blackman, 1963). Both islands were purchased from the native inhabitants by Henry Jacobs Falkinburg in February 1698 and surveyed by Daniel Leeds (Blackman, 1963). Henry Jacobs Falkinburg was acquainted with Indian customs, languages, and for years was friends with the native population. In the past he was an Indian interpreter for early settlers and most likely a surveyor. Other historic houses in the area were the Osborne house on Osborne Island and the Wells house on Wells Island (Fig. 1).

In the Mullica River (Great Bay) area, a Revolutionary War engagement took place between American and British forces in October 1778 (Kemp, 1966). The main objective was Batsto and the privateering center at Chestnut Neck. The British vessels never reached Batsto because the Mullica River was too shallow with sand bars, but they succeeded in burning Chestnut Neck and associated privateering vessels. During part of the military activities the British forces landed at Graveling Point and marched on American forces and massacred 55+/- without any quarter.

They then retreated back to their ship learning of Count Casmir Pulaski Legions at the Willits/Andrews Farmstead (Site 14, Fig. 1). They embarked from Graveling Point. Artifacts from the engagement were found around the Osborne family house (Site 9; Fig. 1) (Stanzeski, 2019).

One of the earliest fish factories in the Mullica Valley is located south of Wells Island in the meadows (Fig. 1, Site 10), starting in 1868 and lasting until 1880 (Able, 2016–2017). The factory processed Atlantic Menhaden used for oils, fertilizer and other products. Remnants of the fish factory can still be seen today.

Understanding Coastal Sites

The Wells Island, Osborne Island, and Graveling Point sites are typical of Late Woodland sites found in central and southern coastal New Jersey. Usually, these sites are located on or near the saltwater meadows. A main-stream or water source can usually be found one hundred to two hundred feet away. A typical camp site may have a wide scattering of oyster or clam shells ranging from fifty feet to three hundred feet adjacent to the main camp site. The predominant shellfish are oysters at most Late Woodland sites. Hard clam are more predominant at Middle Woodland sites.

The size and number of the shell middens varies from one location to another. The larger shell middens are probably sites that had been revisited season after season by the same family or groups of families. Other variables that may have contributed to the varying sizes of the middens include the size of the gathering group, the frequency of seasonal visitations, and the quality of the harvest. Year-round occupations of some sites is possible. The purpose of the harvest was daily use or processing for future use. Shell fish can be dried in the sun and has natural salts to help in the preservation.

Some of the clam and oyster middens are in meadows, away from a base camp. An example would be the Jillson Farm (Site 12) and the Penella Site (Site 13), which were base camps, and the Tuckerton Mound (site 11), the midden. This may be because it is easier to transport the finished product, oyster or clam meat, rather than transport the whole bivalve. The middens may be a main shucking area deliberately located on the outer fringe of the encampment because of the smell and odor generated by the shell debris. Another reason why the sites are found in the meadows or at low tide may be that over time sea level rise has inundated the sites, this being typical of the coast of New Jersey (Horton et al., 2013; Kemp et al., 2013), nearby Atlantic City (Miller et al., 2013) and the Great Bay-Mullica

River estuary (Able, 2020). Evidence of sites found at lower elevation are Site 6, Graveling Point (Site 3; Fig. 1) and the Tuckerton Mound (Site 11).

The inundated sites below the meadows are found in the building of lagoons and in ditches (Sites 4, 5 and 2). Three sites in the area are on high ground including Sites 1, 7 and 8. The changing coast line and the development in the Mystic Island area has occurred not only during the last hundred years (Fig. 1) but also during longer past time periods looking at dates from shell middens below the meadows. Both Wells Island 650+/- B.P., and Steel Site in Cape County 560+/- B.P. (Stanzeski, 1996) were 2 to 3 feet below the surface. The nearby Tuckerton Shell Mound (Site 11, Fig. 1) date was 1530+/- taken 8 feet below the surface. This would be approximately a 7.5 feet average. Another date below meadows, but earlier, was on the remains of cedar tree roots. Near Absecon Creek in Atlantic County taken at 18.3 feet with a date of 4100 +/- B.P. (Ferel, pers. comm.). The above dates would equal 2.5 +/- feet of sediments for ever 500 years. So, in the last 500 years the sea level has risen 2.5 feet +/-. Thus, the depth of sediments on a site can help give an approximate date on the site. More data is needed to show a correlation of depth of sediments to time. But this idea could aid in future work along the shore.

The percentage of lithic material, pottery and fire-cracked rock on most Late Woodland coastal sites is very low compared to other sites along the Delaware River drainage basin. This lack of substantial lithic material is due to the fact that there is very little lithic resource material available locally. There are no quarry sites and even pebble material is lacking in the central New Jersey coastal region. Prehistoric people in this shore area probably turned to shell and bone for tool making material. There is evidence of this material adaptation found at the nearby Pennella site (Site, 13, Fig. 1) (Stanzeski, in review). Both bone and shell were found, which had been used as scrapers, knives, awls, and possible projectile points. Clam fragments found at Late Woodland sites, where oyster shells predominate, could have been disposable tools used to clean the oyster meat. The hard clam, when broken, produces a very sharp edge used as a knife or scraper (Stanzeski, 1981; Brett 1974).

Pottery at Late Woodland sites is as lacking as other diagnostic material. Most pottery is fabric impressed or of Riggins like type. The temper is grit, sand, quartz, or shell. The lack of pottery at sites can best be explained by the area's lack of good sources of clay.

Along the southern New Jersey coast there are other locations that are similar to the complex of sites found at Mystic Island. At the West Creek area there are five sites in one small area (Stanzeski, 1986). Other locations are Port Republic (four sites), Beesly Point (five sites) and Tuckerton area (six sites) (Stanzeski, 1996). The Tuckerton area is interesting because there are some sites that are away from the bay and may suggest a winter site.

Description of a Late Woodland Burial

The following is a description of a Late Woodland burial feature where faunal, flora, shell remains and more were gathered, giving a better picture of Late Woodland period.

In November 1986, on Osborne Island (Fig. 5) (28OC76, Site 7) under a remit oyster midden area, a burial feature was found and referred to as Feature 1. This was part of a stage I cultural Resource Survey (Wilson, 1986). This is in the area of a golf course built in the 1970s. The feature was dated utilizing the C14 method analyzing pieces of carbonized wood and oyster shell. The date of the wood was 400 +/-70 B.P. and the oyster shell date was 720 +/-70 B.P. The C14 dates are both in the Late Woodland time frame.

In Excavation Unit 1 of this site, Feature 1 (Fig. 5) was a flat-bottomed feature with sloping walls, roughly 62 inches at the top and 49 inches at the bottom. The feature originated from the house construction plow zone that was 0–8 inches deep and consisted of a deep black sandy loam with fragments of oyster shell. The feature extended into Level 2 of the excavated unit, which consisted of a yellowish-brown sandy loam. The feature ended when it intersected the top of Level 3, approximately 26 inches from the surface. Level 3 consisted of a yellowish sandy gravel material, feature profile (Fig 5).

The feature had two episodes of fill. Level 1 of the feature contained one and a half bushels of whole oyster shells, and in some parts of the level the oyster shell had melded into a solid mass. Ninety-five percent of the oysters found were bed oysters that occur on mixed muddy sand (Kent, 1988). The oldest oyster was 8 to 10 years old. The average age of the oysters was 6 to 8 years old. There were small amounts of Atlantic ribbed mussel shell fragments, hard clam fragments, common Atlantic slipper shell (young), barnacle fragments and small bone fragments. No cultural material was found except for clam fragments that could possibly have been tools. Level 2 and Level 2A of the feature contained a black sandy loam. Level 2 was the level containing the burial, Level 2A was below the burial. Level 2 and Level 2A had a half-bushel of oyster shells in each. The oysters were 6 to 8 years old.

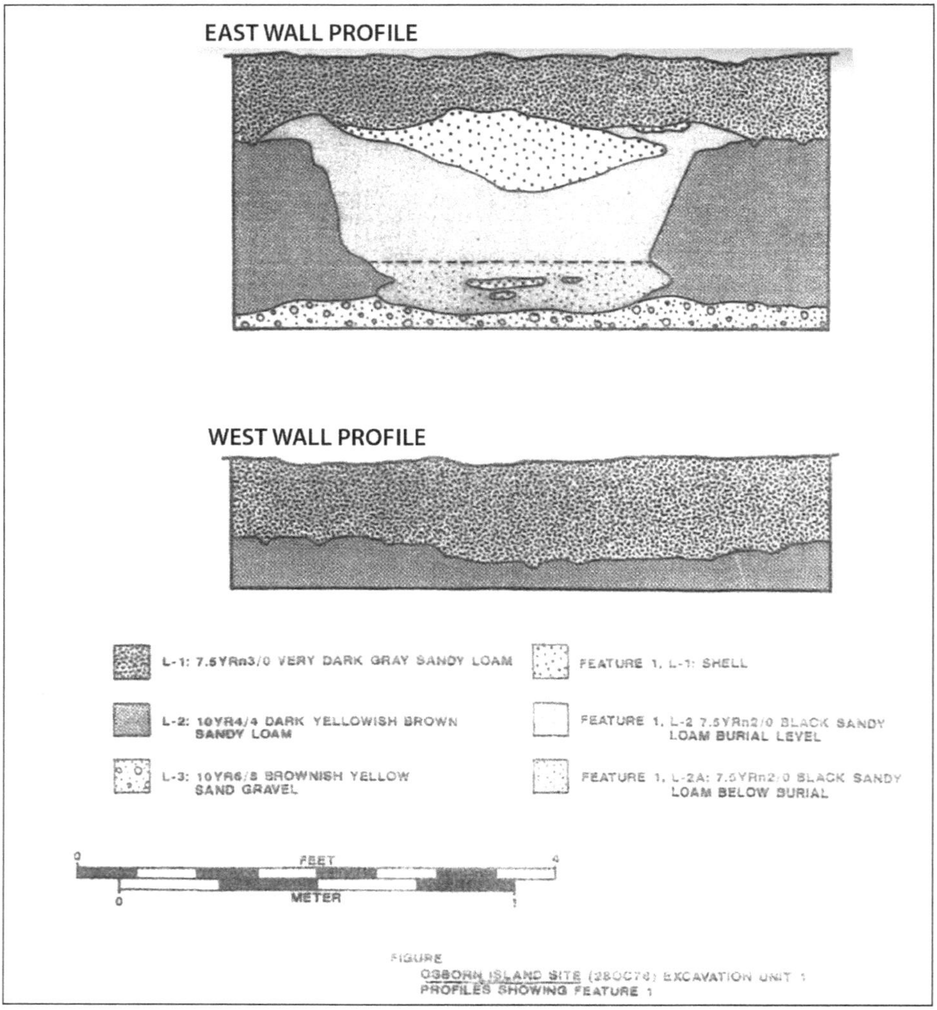

Figure 5 Diagram of Osborne Island site for Excavation Unit 1 with Profiles of Feature 1.

In Level 2, a skeleton of an adult male lay in a semi-flexed position, facing southwest, skull in the southeast, and feet in the northwest. The right arm lay at a right angle across the chest, dislocated at the elbow. The left arm extended downward, the hand touching the feet. The right leg was not completely present, and the left leg was drawn up loosely to the body. In October 1988 the remains were delivered to the Smithsonian Institution for analysis by Dr. Douglas H. Ubelaker. The remains submitted from New Jersey originated from one very elderly adult male (55 to 70 years old) approximately 5.7 feet in stature. The man suffered from advanced osteoporosis, degenerative joint disease (including a collapsed twelfth thoracic vertebra) and severe dental disease (Ubelaker, 1996). Inside of the stomach cavity of the burial there were two clusters of very small fish remains and seeds. In 1992 the burial remains were reinterred; there was an Indian ceremony, which was presided over by Delaware Indian representatives James Revey and Richard Gilbert.

In Level 2 there were small amounts of Atlantic ribbed mussel fragments, clam fragments, common blue mussel fragments, Atlantic slipper shell (young), a good amount of snail shells and barnacle fragments. Over 300 bone fragments of deer, turtle, fish, and bird were found in Level 2. There was also a small amount of fish scales, seeds, and nut fragments recovered.

In Level 2A the shell remains were the same as Level 2, with some bone. The cultural material included seven small potsherds, the same quartz and argillite temper potsherd types found in Level 2. One possible oyster shell tool was found on the feature floor, and hard clam shell fragments (possible tools) were recovered from Level 2A.

The 26 potsherds (12 were small fragments) were in Level 1, 2, and 2A. The types found were shell tempered smooth on the outside and fabric impressed inside, quartz temper fabric impressed outside and smooth inside, chert tempered fabric smoothed over inside and outside (Riggins like). The sherds were from three different pottery vessels and were part of the fill found around the burial. All lithic material was found in Level 2. The material was 5 chert and 3 jasper, all pebble material. The flakes were decortification, late stage and thinning flakes. The tools found were 1 jasper flake tool fragment and a possible chert distal fragment.

A total of 26 clam fragments were found in all three levels. The clam shell fragments found throughout the excavated unit and the feature could have been tools or the byproduct of making tools, possibly used for cleaning the meat from the oysters. No whole clamshells were found. Clamshell maintains a sharper cutting edge than oyster shells. Tribes in South America have been observed using clamshell as a tool. The Yahgan of Tierra Del Fuego were observed exhibiting "even today when metal tools are readily available a marked tendency to use anything which comes to hand, be it a shell fragment or a flint chip" (Lothrop, 1928). The clam shell fragments found throughout most of the Late Woodland sites are most likely tools. One possible oyster tool was on the bottom of the feature. The ventral end of the oyster shell was worn up to the adductor mussel scar. The shell could have been one of the tools used to dig the feature. It was imbedded in the yellowish sand and gravel below the burial and was most likely present on the site before the feature was used for burial.

Two post holes were located, one in the southeast corner of the excavation unit and one in the bottom of the feature on the northeast side. The feature had most likely been used as a storage pit. The sand and small gravels below the burial (Excavation Unit, Level 3, Fig. 3) would have provided good drainage if water had gotten into the feature. These sands would have helped to prevent spoilage of material stored in the feature. Post holes show the position of possible supports for a cover for the storage pit feature. The feature was then used as a burial grave and Level 2 and Level 2A was the fill that accompanied the burial. The remains of marine molluscs and other animal bone, in both Level 2 and Level 2A, could have been ordinary refuse. Alternatively, a small feast for the individual had taken place during the burial and the remains interred in the grave.

The younger oyster (spat oyster) and younger slipper shell in Level 2 and 2A suggest that the buried individual was interred around the end of the summer or early fall, assuming the bivalve spawning took place in the spring or early summer and there were favorable growing conditions for the young oyster and slipper shell. The younger oyster, bone, lithics, pottery, seeds and nut fragments in L-2 and L-2A, with less to none in Level 1, suggest that Level 1 was a different episode of fill placed on the burial at a later time.

Faunal Remains from Osborne Island Burial

The following is a review of the animal bone associated with the burial located on Osborne Island (Site 7, Fig. 1). A total of 300 bones were found and most were identified. The highest numbers of bones (n=170) found were from large birds. Identifiable bones included turkey (4), snow goose (7), and Canada goose (1). A large number of shafts, humerus, and a smaller number of tibiotarsis and a tarsmetatarus were broken of at the ends. At first the bones were thought to be beads because both ends were broken off. The ends of the bone were broken off to extract the marrow from the shaft. Some of the bones showed cut marks, puncture marks and crush marks. The second highest numbers of bones (n=165) were from fish. Identifiable bones were from sea bass and bluefish. Also, one possible tuna size vertebra was found. The third highest numbers of bones were from mammals, with the most being of deer (n=15). Some of the deer bones showed cut marks.

The other mammals were meadow vole (n=5), gray fox (n=1), striped skunk (n=1), short tail shrew (n=3), least shrew (n=2), and river otter (n=1). Also, other fragments (n=46) of mammal bone could not be identified.

The types of reptiles and amphibians found were diamondback terrapin (n=12 fragments), followed by eastern box turtle (n=7), spotted turtle (n=6), snapper (n=1), red bellied turtle (n=1), and racer snake (n=1). The amphibians were two bones from a Fowler's toad. The eastern box turtle bones were most of the carapace fragments that showed cutting, scraping, and residue. The carapace of the turtle was most likely used as a bowl.

The assemblage of bone was different than most prehistoric sites in that bird bone outnumbered deer bones. In most prehistoric features deer bones are the highest number found. Food processing of the bone was indicated by cut marks on both the deer and bird bone. All soil from the stomach cavity area of the burial was gone through with a dental pick and floated because of the texture of the soil. The soil was a dark black with a grease type feel. In the stomach cavity, small seeds were found throughout with two batches of tiny fish bones, in one area. These were likely part of the individual's last meal.

In the burial were also only one bone from 5 different types of animals, fish, and reptile. A vertebra from a possible tuna size fish, racer snake (posterior skull), river otter (left carnassial tooth, gray fox (tooth) and striped skunk (left denture), and one claw core was found but could not be identified as to type of animal. There could be an importance to each of these bones. If the tuna had been eaten there would be more bones from such a large fish. The other individual bones were either teeth or skull fragments. Again, there were no related bones in the burial for these five (representative) animals. A skunk would not be particularly edible. Fox teeth, with no other bones of a fox, have been found in Late Woodland Monongahela burials in Pennsylvania (M. Stewart, pers. comm.) and by Butler (1971). Bear canines were found with Susquehanna burials (Kent, 1984). The importance of the teeth has been seen through time. The other teeth, vertebra and claw core were of possible importance to the individual buried or to the people that buried the individual. The five different bones could also be a medicine man bundle.

The season of the burial was likely late summer or fall because of the large number of migratory bird bones found and the presence of young spat oysters. The number of bones and the different type of animals suggests that at the time of the burial a possible reception occurred, as in burials in the Upper Delaware (Kraft, 1973). Noteworthy was that one leg of the buried individual was missing.

Conclusions

Research for this paper began years ago, in the hope of understanding the prehistoric use of the New Jersey shore. The Mystic Island area was only part of the research. The area shows several sites with the mean occupation being Late Woodland. The information from the burial site gives a picture of a feast of the dead, a medicine bundle present, and the individual's last meal. The two C-14 dates show a difference in the use of shell and wood carbon for dating. But still either can aid in dating. Curation of shellfish was the main activity, leaving large to small shell middens thoughout the area. More than half the sites are under the meadows and only found during lagoon cuts, in mosquito ditches, or low tide. The rising sea level on Late Woodland sites has happened with an estimate of 500 +/- years equals 2.5 feet +/- of increase along the New Jersey coast. The depth of bottom sediments found on (or above) a site can give an approximate date to the inundation of the site. The sites and all prehistoric time periods in the area have less lithics and pottery than most sites found on the Delaware River drainage. The combination of less artifacts, rising sea level, and development makes this a challenge for archaeological work in the area.

Acknowledgments

This paper would not have been possible without the first work done on Osborne Island for Charles I. "Budd" Wilson Jr. in 1986, a legend in New Jersey Archaeology. Subsequently, Dave Parris and Louise Miller provided faunal identification. I would like to thank Louise Miller and Dave Parris for the bone identification and Douglas Ubelaker for the human remains. The four C-14 dates were funded by the Archaeological Society of New Jersey. Andy Massler provided some of the site locations and artifacts in the report. Charles Bello and Leslie Eisenberg provided a picture of the gorget. Photos were done by John Stanzeski. Dr. Kenneth Able provided comments on an earlier version of the manuscript.

References

Able, K. W. (2016–2017). Manufacturing from menhaden: A history in the Mullica Valley. *Sojourn 1*(2): 75–82.

Able, K. W. (2020). *Beneath the surface: Understanding nature in the Mullica*

Valley Estuary. Rutgers University Press.
Bello, C. A. & Eisenberg, L. E. (1988). An analysis of the skeletal and artifactual remains from a disturbed coastal site in Ocean County, New Jersey. *Bulletin of the Archaeological Society of New Jersey,* No 43: 45–51.
Blackman, L. (1963). *History of Egg Harbor Township, Burlington County New Jersey*. Trenton Printing Company, John Mathis Foundation.
Brett, P. A. (1974). The Maurice River shell tool complex. *Man in the Northeast, 7*.
Butler, M. (1971). Three archaeological sites in Somerset County, Pennsylvania. *Anthropological Series of the Pennsylvania Historical and Museum Commission, 1*, 509–43.
Cross, D. (1941). Archaeology of New Jersey. *Archaeological Society of New Jersey and New Jersey State Museum Trenton, 1*.
Goos, N. R. (2011). *Atlantic County Historical Society and Geneological Journal, 16*(4), 58–68.
Horton, B. P., Englehart, S. E., Hill, D. F., Kemp, A. C., Nikitina, D., Miller, K. G., & Peltier,W. R. (2013). Influence of tidal-range change and sediment compaction on Holocene relative sea-level change in New Jersey, USA. *Journal of Quaternary Science 28*, 403–11.
Kemp, F. W., Horton, B. P., Vane, C. H., Bernhardt, C. E., Corbett, D. C., Englehart, S. E., Anisfeld, S. C., Parnell, A. C., & Cahill, N. (2013). Sea-level Change During the Last 2500 Years in New Jersey, USA. *Quaternary Science Reviews, 81*, 90–104.
Kemp, F. W. (1966). *A nest of rebel pirates*. Batsto Citizens Committee.
Kent, B. C. (1984). *Susquehanna's Indian Anthropological Series of the Pennsylvania Historical and Museum Commission, 6*, 176.
Kent, B. W. (1988). Making dead Oyster talk. Maryland Historical Trust.
Kraft, H. C. (1973). Feast of the dead. *Bulletin of the Archaeological Society of New Jersey, 31*, 29.
Lothrop, S. K. (1928). *The Indians of Tierra Del Fuego*. New York Museum Of American Indian, Hege Foundation.
Miller, K. G., Koop, R. E., Horton, B. P., Browning, J. V. & Kemp, A. C. (2013). A geological perspective on sea-level rise and its impacts along the U.S. mid-Atlantic coast. *Earth's Future, 1*, 3–18. doi:10.1002/2013EF000135.
Mounier, R. M. (1999). A brief history of investigation of the Tuckerton Shell Mound. *Bulletin of the Archaeological Society of New Jersey, 54*, 43–44.
Parris, D. C., Gallagher, W. B., Grndstff, B. S., & Tample, C.D. (1988). Quaternary mammals from the continental shelf of New Jersey. *The*

Mosasaur, 4, 101–10.
Ritchie, W. A. (1969). *The archaeology of Martha's Vineyard*. The Natural History Press.
Skinner, A., & Schrabisch, M. (1913). A preliminary report of the archaeological survey of the State of Jersey. *Geological Survey of New Jersey Trenton Bulletin, 9.*
Stanzeski, A. (1981). Quahog "shell tools." *Bulletin of the Archaeological Society of New Jersey, 37*: 15–18.
Stanzeski, A. (1996). Two decades of radiocarbon dating from the New Jersey shore. *Bulletin of the Archaeological Society of New Jersey, 51*: 42–45.
Stanzeski, A. (2019). McDermott site, Martha's Vineyard. *Bulletin of the Massachusetts Archaeological Society, 80*(1–2): 55–68.
Stanzeski, A. (2019). Chain on Osborn Island, a bad day in the American Revolution. *Bulletin of the Archaeological Society Of New Jersey, 74*, 115–20.
Stanzeski, A. (2001). The Tuckerton Shell Mound. *Bulletin of the Archaeological Society of New Jersey, 56*, 47–50.
Stanzeski, A. (In review). Archaeological perspectives on the New Jersey Shore. *Bulletin of the Archaeological Society of New Jersey.*
Ubelaker, D. H. (1996). Aboriginal human remains from Osborn Island. *New Jersey Bulletin of the Archaeological Society of New Jersey, 51*: 73–74.
Wilson, C. I. (1986). Archaeological investigations at Osborne Island. Report Addendum, Block 201, Lot 65, Little Egg Harbor Township, Ocean County, New Jersey. Report on file at the Historic Preservation Office (NJDEP), Trenton.

Tables

Table 1: Faunal Remains from Level 2 and 2A in Osborne Island Burial Site

Common Name	Number	Scientific Name
Short-tailed shrew	3	*Blarina brevicauda*
White-tailed deer	15	*Odocoileus virginianus*
River otter	1	*Lontra canadensis*
Meadow vole	5	*Microtus pennsylvanicus*
Dog or gray fox	1	(dog) *Urocyon*

		unereoagenteus
		(Fox) *Urocyon cinereoagenteus*
Striped skunk	1	*Mephitis mephitis*
Least shrew	2	*Cryptotis parva*
Wild turkey	4	*Maeleagris gallopavo*
Canada goose	1	*Branta canadensis*
Snow goose	7	*Chen hyperboria*
Spotted turtle	6	*Clemmys guttata*
Racer (snake)	1	*Coluber constrictor*
Fowler's toad	2	*Bufo woodhousei fowleri*
Eastern box turtle	7	*Terrapena carolina*
Diamondback terrapin	12	*Malaclemys terrapin*
Tuna (large)	1	*Scombridae*
Sea bass or bluefish	165	*Serranidaw / Pomatomidae*
Large fish framents		
Small fish (stomach cavity)	2 whole	
Small fish fragments		
Claw core	1	
Mammal	46	
Bird	170	
Total	300 +	

Appendix 2

Time Well Spent: Fishing Along Seven Bridges Road

Steve Spahn

The following stories were gleaned from experiences and observations while fishing on the marsh along Seven Bridges Road over the past 25 years. We moved to southeast New Jersey in 1998—new job, new schools, new house. I discovered the marsh while we were still living in the motel. As work and family allowed, I could be found on the marsh at almost any time of the day or night—before work, after work, in the midnight hour. I was hooked. But the funny thing is the fishing is less about the catching than it is about immersing oneself in the mind-boggling variety of life that abounds on the marsh and in the waters that sustain it, the cycles and patterns of life, death and regeneration.

From Mud to Mud

The marsh in southeast New Jersey stretches for miles sandwiched between the barrier island glitter towns and the eastern edge of the New Jersey Pine Barrens. Everything in the marsh rests on the bosom of mud. It is the mother nurturer of the insects that hatch out every year in numbers so colossal as to make the national debt look trivial. In early spring, mosquitos and biting gnats hatch from the pools and puddles that dot the breast of the marsh. Greenheads, large biting flies, live in the mud as predacious larvae, eating other insects and small crustaceans. In June they rise from the mud in marauding swarms to form, with the gnats and mosquitos, the South Jersey Parasitic Pantheon, the Father, Son and the Holy Ghost of evolved parasitic perfection searching for warm blood, namely mine, in order to continue their breeding ritual. Gnats and mosquitos are most active during the dark hours. Greenheads are sight feeders—large powerful fliers like horse flies on steroids. But during transitions (dawn and dusk) all three operate with evolutionary efficiency. Gnats clog any

The Fish Factory, rusting away on Crab Island behind Brigantine and Atlantic City. Built around 1930, it serviced the industrial harvesting of bunker (aka menhaden or mossbunker), a ubiquitous forage fish along the East Coast. Everything eats bunker, from striped bass to bluefish to humpback whales. Everything. It was abandoned in the 1960s. In the dark of night, some say a ghostly glow can be seen from the now cold fires that rendered millions of bunker into fertilizer, cat food, and ingredients for paint and nail polish. Maybe it is just wisps of sea fog playing in the ribs of the decaying buildings. I don't know.

exposed orifice—eyes, ears, nose—mosquitos swarm in choking clouds while greenheads bombard their prey with audible 'thuds'.

The marsh is a predator's paradise. Everything out here bites, stings, sucks, rips, or rends. Swallows, swifts, and martins carve the sky into ephemeral parabolas, picking off insects that rise in clouds from the marsh. Heron and egret ring the live holes making life miserable (and short) for the killies and silversides stranded when the sea retreats. The gulls pick at whatever they can get their beaks around, from rotted fish to live clam, mussel, or crab which they drop onto the macadam of the road, breaking open the shell for the meat inside. At night or in the low light hours, skimmers work against the tide lines, their lower beak trawling the water as they hunt for anchovies. Ospreys patrol the open channels from high in the air, ready to plummet into the water in a breathtaking, bent wing dive onto the back of an unwary fish. Cormorant, loon, and even seal hunt beneath the water, each in their own season. Spider crab and blue claw fight over the submerged real estate while armies of fiddler crab parade along the tidal zone waving their one, oversized claw in a comical show of bravado. Eagle, snowy owl, fox, or feral cat will pounce on an unlucky duck or goose leaving scatters of feathers and feet. In my eyes the marshlands are like a wilderness in a bottle, squeezed into an eminently uninhabitable stretch of mud, tidal pools and marsh grass between the Garden State Parkway and the bright lights of the Atlantic City casinos. The marshlands are at once a perfect wilderness yet flawed with the human presence. One relic of the human presence is the Fish Factory whose story is captured with the picture above, taken from the marsh along Seven Bridges Road.

Night Trips

There are five bridges to cross on the marsh road. That's why the locals call it Seven Bridges Road, I guess. The last two bridges are narrow, single lane wood plank affairs. Traffic is controlled by lights at either end of the bridges. Also, at either end of both of these bridges are signs which read:

Signs and traffic lights be damned, in years past, when fish were running, the last bridge, on a weekend summer night, the one over Little Sheepshead Creek, could be plugged with people. Crawling along in first gear, I do my best to avoid crushing any of them beneath the front bumper. More importantly I am on high alert for a mis-timed back cast and a bank sinker in the windscreen. I squeeze the car between the bodies dripping a toxic brew of bug juice and metabolized lite beer. Sometimes I shake my head and chuckle at all the gear they lug out onto the bridge: tackle boxes, boom boxes, six packs, coolers, lawn chairs, crab traps, grapples, portable generators connected to lights hung over the water to attract bait. The list could go on and on. It's like your local Walmart collapsed into one aisle, like a black hole of merchandising. In the morning they'll all be gone, replaced by seagulls squabbling over the trash they leave behind.

The harshness of the lights and the sound of the portable generators set against the velvet dark matter and primal silence of the marsh give the bridge scene, on a summer night, a surreal glow playing bookend to the gash of light from the Atlantic City casinos ten miles distant on the southeast horizon.

The road ends about a mile past the Little Sheepshead Bridge. The asphalt just stops and the road turns into a series of gravel and sand pot holes maybe 40 yards long. The area is walled on three sides by scruffy elderberry and spartina or big cordgrass. There are no lights. Rickety wooden utility poles veer off over the marsh to the Rutgers marine lab now occupying the old Coast Guard station. An elevated wood plank vehicular causeway runs over the marsh to the marine lab through a chain link gate. Even ricketier poles strike off through the marsh to the north fading from sight into the horizon. They are still strung and look intact, but I cannot see where they go or what they do other than provide nesting opportunities for osprey and eagles.

At the far end of the lot at the road's end a sandy path cuts straight through a dense thicket of brush, scrub cedar, elderberry trees, and poison ivy leading to the edge of the bay. The elderberry bushes are dense and tough, the branches gray and weathered from the maritime climate. In the spring and summer, red-winged blackbirds use that cover to defend their nests from raiding gulls. In late summer and early fall, migrating monarch butterflies weave their drunken-like way back to the highland pine forests of central Mexico, stopping briefly for rest, nourishment, and shelter before they lift again, propelled on their journey by the northers the season brings.

The parking area at night is a spooky place especially for those of us who are afraid of the dark. Without the reassuring sound of the engine and the stab of the headlights the silence is primal and the darkness elemental. Primal silence does not mean quiet but there is not a sound that is out of place: the wind moans in the wires and hisses through the brush; a jet aircraft high overhead whines its lonely trail through the heavens. The rush of the sea against the bank congeals the fundamental elements of the night world at the end of Seven Bridges Road: water, mud, sod, and dark matter.

Pulling into the area at night it's good to see who is already there. You get to know the cars more than the guys who drive them. The best is no cars as is often the case. Second best is cars you recognize. I know that if I ever have a problem it's going to be in the lot. That is if I don't break my leg in a sink hole, step in deep mud, get sucked by a rancid mosquito, bury a hook in my hand landing a bass, or step on a cow nosed ray when they cruise the flats in late summer. In my risk analysis I can suffer these as the fortunes of my nocturnal trysts with Mother Nature. But the parking area has me more concerned with thoughts of the forces of Muthas Unnatural: cars filled with misdirected youth, or members of the local Hell's Angels cartel carousing in the lot. Have I ever come close to a problem? No. To be sure, there have been teenagers in cars. But they were too busy trading hands, slapping bugs, and solving mysteries I can no longer recall than to make sport with a grayed out Swamp Rat backing along the bushes with his hand gaff at the ready.

I have my gear, stuff I need, and then I have other stuff, gear that is discretionary and usually a pain in the ass. I move when I work the sod banks or the flats so the lighter the better. I dress in the dark allowing my eyes to adjust. Waders go on first. I cinch the waders at the waist not just to slow the rush of water if I take that one step too many, but it lets me stow a box of flies or plugs inside my shirt so they don't slide down to my feet. In that way, too, I avoid having to carry a ditty bag on my shoulder. Fishing vest goes on. Check the pockets for essentials like bug juice and leader. I drape a pencil light on a lanyard around my neck and check the beam. Clip a pair of long nosed pliers to the vest. Find the hand gaff and clip the tether to one of the rings on the vest and hang it by the hook to the same ring. The gaff is simply to immobilize the fish by slipping the point through the membrane in the lower jaw. I have courted hook in the hand disaster for too many years than to tempt this piece of my karma. Trying to get a handle on a struggling fish with hooks flying around standing in the water up to my waist in the dark has gotten way beyond my calculus to tempt fate. But I know Fate is out there, ready to swoop down at the first opportunity—BLAM!—like a hawk on a little furry bunny. As Taj Mahal tells us in *Good Mornin' Miss Brown*: "Man if you ain't scared, you ain't right!"

The marsh receives me, wraithlike: Walking out, through the grass and the mud, past dead pools and live tidal channels, I feel the stars shine through me. In the moonlight I become one with my shadow. No. It does not fool the bugs, not for one wingbeat. I don't use the light. If I'm going to hit the usual spots I pretty much know where I am going, one step after the other. The sensory world remains, but touch, sound, and smell define the boundaries. On a dark night, the visible world shrinks to one step at a time. But tonight, moonlight lays like a silver blanket over the expanse of mono-toned marsh grass. The dark water refracts the brilliance into diamond-like shards. On cloudless nights, the full moon is too harsh. I don't like it. I have never had a good night on a cloudless full moon. But the moon shining through some high cirrus transforms the marsh into a magical happening. Even though this is New Jersey, there are few other primates out here in the dark. Most times it's just me, the token primate. But the bugs are here and they are not fussy.

I step into the water at the head of the flat near the marine lab. Two feet from where I stand, to my right, the flat drops in one step into 40 feet of water. But I know that as I move northeast it widens to over a hundred yards across, the sod banks to my left and the edge to my right. The outgoing tide tugs at my thighs. I am now, I think, truly invisible. The distant light from the Atlantic City sin palaces taint the fabric of the night. No, they can't see me I think. But I sure as shit see them: the Borgata, the Trump, the Tropicana, all the icons of discretionary decadence fading off to the southeast in a fuzzy glow. They are always the same, every night.

I turn my back on the lights and work northeast along the flat. I'm using a nine-foot ten weight loaded with an 11 weight intermediate line finished with a 7' leader tapered to a 16 lb tippet. I don't need to make particularly long casts. Too much line is a hazard in the dark anyway. Lashed to the leader is a fly I tied for the dark hours: spun purple deer hair head trimmed really tight with two big yellow eyes cemented on a black and purple Polar Fiber body, a couple of purple Chinese saddle hackles on each side with some Flashabou and Krystal Flash thrown in.

The flat broadens and soon I am yards from the sod bank, up to my waist in the black, moon flecked water. The tide is about half gone and moving nicely. To fish this flat I need a dropping tide. The incoming runs like a freight train through here and it is not for wading fishermen. On some nights there is phosphorescence in the water but not much tonight. A couple of skimmers come careening out of the dark flying up tide only inches above the surface, their lower beak in the water trawling for silversides. They are as surprised as I am that we have found each other's company. The bugs are not bad with the south breeze. South wind is my least favorite for fishing this part of the swamp but Swamp Rats take what wind they can get when it comes to bugs. The south wind is in my face coming off the water and it makes the flat noisy. There doesn't seem to be much doing as I continue to work northeast. The repetitive motion of casting soon has me in that zone where time falls away and my mind floats like a butterfly collecting the nectar of the universe.

I have no action and I have heard nothing resembling feeding fish. That's one of the great things about working the long rod especially in the dark: one can go from blissful abstraction to a fiercely focused state of being in an instant. Just add a fish and the chaos will follow! But tonight, Orion yawns in his mesmerizing passage overhead. The sensual caresses of the black water and the warm summer breeze have seduced me with their sirens' song and I decide to wade out to the shelf so I can work a cute little hump right on the edge of the drop off. It is not without angst that I do this. I must admit, I am spooked by the proximity of deep water. It's not that I have a problem with deep water. Maybe it's just the circumstances of having waders on. Maybe it's the darkness. Maybe it's because I have blundered over the edge of this flat a couple of times, not keeping track of where I was. No harm done as I spluttered my way back onto the shallows none the worse for wear other than a red face and the sloshing noise coming from my waders walking back to the parking area.

Suffice to say I outlived the bravado of my younger days. At least I think it's dead. Most everything else is. I have often wondered why the hubris of youth is replaced by the paranoic caution of old age. Is it that at some point, having been pummeled, nicked, and generally pushed around by life, we realize that we are not going to live forever? Or is it diminishing capability and capacity that sobers us up? So the point of our life becomes simply keeping it alive. Not Stayin' Alive like the B-Gees. Keeping alive. When does lean and mean turn into skinny and scared?

You can tell, sort of, where the drop off is from the tide lines and upwellings traced on the surface of the water. Sort of. But it's not like a stake in the mud with a sign saying, " DROP OFF HERE." It just indicates some proximity. In the dark it's dicey. One step too many puts you into 40 feet of water. I also have a pretty good idea judging by my distance from the bank in this particular spot. My heart rate also goes up, exponential to my proximity to the deep water.

I gingerly edge my boot forward and find it suspended in the open Atlantic! 'Boom! Boom!' Whew! Yup, I found it! In the darkness the rush is for real. I am now only inches away from water deep enough to float anything from Jaws to the Queen Mary! The outgoing tide presses in on the ledge in this area, kind of a hollow or cove on the edge of the flat and forces itself up and over this little hump. Boils and eddies mark the surface and drift down tide, drawing and re-drawing ephemeral sketches of the fluid mechanics of the place. It has always looked to me like a classic spot for bait to get ambushed. Not that I have had spectacular success here, but more, I believe in it. I really know this small place and it is as if it knows me and for some small passage of time, here in the deep of night, I truly can be a part of the Great Becoming all around me. That's why I am here, I think.

The great fin ripped the surface about a rod's length from where I stand shredding whatever delusions I may have had concerning the oneness of life. The body carrying that fin breached, visible as a gray splotched, glistening hulk so close I can see, even in the moonlight, purple tracings of scars on the creature's back. Everyone knows the fight or flight reaction, those infinitesimal dollops of time that drip by in slow motion as we take stock of an unexpected, threatening situation and you freeze as if pole-axed, your heart in your mouth. In a second or two the water healed itself and the south wind pushed wavelets over the hole where the fin had been.

I am as naked as it gets: flight impossible because my feet are planted below sea level in the ooze of the flat on the edge of the drop off and I am 100 yards from Mother Earth. OK, sure, it was a transiting bottlenose dolphin. I heard the creature roll and blow again some seconds later, down tide. But then who said they all have to be Flipper clones. And they are big—600 pounds if they are an ounce. If they are as smart as they're supposed to be there's got to be a wise ass or maybe even a dolphin variant of a psychopath out there in the sea somewhere. Who knows. Another one blows, further out and slightly up tide. My hands still shake but I am breathing somewhat normally again. I look up. Orion has journeyed further along the dome of the sky. It must be about 2 AM by now. I free my boots from the muck of the flat and move slowly inshore toward the sod bank and the parking area, still searching for my part in the Great Becoming.

How Deep Does It Get

In the 2000s, like sometime from mid-April to mid-May, big spawning weakfish would move into Little Sheepshead Creek, spreading out around a hole on the Fish Factory side of the bridge. The best fishing was from the banks of the creek. In the spring, the Bridge People abandon the Little Sheepshead Creek bridge for the sod banks on the south side of the creek. It's mostly the same old Bridge People except they do it on the marsh. As a swamp rat, I am not particularly fond of this fishing. It violates almost every fishing principle I have come to embrace as a lifelong devotee of this blood sport called fishing: avoid crowds, fishing contests, gang bang fishing, and light beer. The marsh wasn't made for gang bang fishing I think. It's like a fishing circus with too many clowns. But it's the best game in town and the fish are beautiful: 5–6 pound weakies were common, 10 pounders caught every year. So every spring I take my Manifesto of Fishing Principles, fold it into many anal creases, stuff it in my shirt pocket, buy a quart of Miller Light and head for Little Sheepshead.

The walk out is easy, too easy. Some community minded hero dropped a couple of timbers across a nasty mud hole running next to the road embankment. The mud is the color of a dirty, iridescent pumpkin. Over these are laid sheets of delaminating plywood probably scavenged from a nearby derelict structure of unknown origin now sinking into the marsh. Once you make it over that wallow, it's damned straightforward. Just a short stroll to the creek with maybe a couple of soft spots but not bad. I did see a big dude get hung up one day on his way out. That spot does not look particularly hazardous. In fact, probably dozens of guys walk through there every week without thinking about the mud. Somehow, this sport finds a soft spot where nobody else could and he's in up to his thighs. I watch the guy from the banks of Little Sheepshead. He's wallowing left and wallowing right and going nowhere fast except probably down; however, not at an alarming rate except maybe to him. So nobody pays him much attention except the ospreys nested atop a dead power pole nearby. Finally, somebody on the way out stops and it looks like they are making progress. But next time I look it seems he has developed a serious list and winds up on his side. At this point they abandon the boots, and he rolls over on his back and out with his boots still in the mud. Swamp rats, even light ones with big feet, always check the mud.

There is one serious wallow. It's not exactly a wallow but the end results are the same. It's a small creek with steep banks and it drains into Little

Sheepshead. Anything higher than mid tide and you can't cross it and stay dry even with chest waders. The bottom is soft, very soft; and deep. This mud is D-E-E-P.

I haul on the long rod, a 9' 10 wt. and I watch, and I wait from the other side of the little creek. It is inevitable. Why do they try to cross it—in deck boots, hip boots, and waders—Why? Because it's there, I guess. I can't figure it out. The fishing is as good on one side as on the other. Why do I not even stop to fish until I'm on the other side? Well, shit, to get away from all of them that can't. It's the psychological necessity to put some barrier between me and the Bridge People. The clientele, I think, is of a higher caliber on the other side.

Invariably, a dude will approach the wallow. If it's low tide, the creek is almost empty of water except for dribbles of psychedelic scum draining from the marsh. The mud lays there like a malevolent chocolate pudding, ready to suck down the wearer of the unwary boot. The sucking down is one thing. In the mud wallows, balance is THE thing, as any swamp rat will tell you. Anyone who has watched female mud wrestling can also attest to that. They're always falling down—so too on Little Sheepshead Creek in the early spring.

I have pondered the physics of walking in deep mud. It is not, I conclude, actually walking. It's about getting somewhere for dubious reason.

The repetitive and soothing motion of the long rod has my mind in the Zen Zone again.

"Hey, how deep does this thing get?" yells a dude in hip boots already in up over his knees, snapping me back to genuine mud.

"Stay to the left," I suggest avoiding the answer because there is no answer! It gets as deep as it friggin' wants! The dude tries to extract his left leg but then his weight shifts to his right foot which of course sinks deeper into the wallow—Presto! After a few wild arm rotations, he is half prone on his right side in the mud and everything is back in balance. I turn back to fishing those big weakies. I know that mud cruising is more than highly tuned muscles with synapses honed to lightning speed. It is more than fine motor skills in the balls of your feet making minute and instantaneous adjustments to the mud. It is, I know, mind over mud.

Appendix 3

Rutgers University Marine Field Station's Role as a Technology Development Hub for Modern Oceanography

Oscar Schofield and Scott Glenn

Center for Ocean Observing Leadership,
Department of Marine and Coastal Sciences,
School of Environmental and Biological Sciences,
Rutgers University

> "I walk into our control room, with its panoply of views of the sea. There are the updated global pictures from the remote sensors on satellites, there are the evolving maps of subsurface variables, there are charts that show the position and status of all our Slocum scientific platforms, and I am satisfied that we are looking at the ocean more intensely and more deeply than anyone anywhere else."
>
> Henry Stommel, "The Slocum Mission" (1989)

Abstract

The Rutgers University Marine Field Station (RUMFS) has been central to many significant technological contributions in ocean sciences over the last three decades. It has been home to the development of many core technologies that are underpinning the development of a global ocean observing system. These ocean observing systems will enable a new era of economic growth and global environmental sustainability critical to thriving on this ocean planet. We will review the evolution of ocean observing technologies, the technological contributions at the Rutgers University Marine Field Station, and comment on the future moving forward.

Introduction

For centuries, society has studied oceans relying on data and observations gathered from ships during cruises of limited duration. This research approach has resulted in major advances in our understanding of global ocean circulation, the energy associated with weather-driven circulation, plate tectonics, global ocean productivity, and climate-ocean coupling. The research successes have expanded our view of Earth and ocean processes, and have demonstrated a need for an improved way to study the ocean using new sampling approaches spanning time and space scales that are not effectively carried out using ships. To address this observational gap, the scientific community has consistently called for the development of the capability to maintain a continuous sampling and monitoring presence in the ocean (National Research Council, 2003; Schofield and Tivey, 2005).

The difficulty in sampling the oceans coherently is one of the central problems for oceanography. It can introduce biases which might compromise our understanding of many ocean processes. For example, the role that large storms can enhance the biological productivity in the nutrient-poor deep ocean was not appreciated until they were sampled by fortuitously timed open-ocean cruises (Glover et al., 1988) and/or the deployment of moorings that provide measurements when ships could not remain at sea (Dickey et al., 1998). This finding was exciting in that sampling in a large storm from a ship can be dangerous and limit when/where scientists can sample. Given this, to realize the goal of continuous sampling required the development of new tools to expand the sampling capabilities.

The field station at the Rutgers University Marine Field Station (RUMFS) has a long storied history in providing ecological insights into coastal ecosystems (Able 2015, 2020). The field station also has another distinguished past that includes being a location that has been central to the development of novel ocean technologies that have been revolutionizing how the ocean is being sampled. We will review the novel observational and modeling tools that were developed and tested at RUMFS. Additionally, we will also highlight a range of societal applications that these technologies and networks are being used to address.

Sampling improvements are needed to expand the ability to incorporate time and space measurements. This is needed as many critical ocean processes span multiple time and space scales. Sampling many time and space scales is difficult using a single type of sampling platform (ship, mooring, etc.) (Figure 1); therefore a comprehensive sampling program will require combining different sampling platforms to span the appropriate time and space scales. Historically since the 1970s, oceanography's sampling programs have relied on the tools available at the time including satellites, moorings and ships. These tools, while highly effective had some limitations. For example, 1) ships can only operate under safe conditions for the human operators, 2) satellites can only directly sample the ocean's surface, and 3) moorings cannot provide spatial data critical to understanding many processes. New tools complement these traditional sampling platforms, and overall a multiplatform approach will need to be adopted to address the critical science questions. RUMFS has been critical as it provided a coastal test-bed in which to innovate new tools. Some of the new technologies developed at the Rutgers University Marine Field Station include:

Ocean Robots: Since the last century, science fiction novels and films have highlighted the potential for autonomous underwater sampling/exploration. The birth of autonomous ocean robotics began with Autonomous Underwater Vehicle (AUV) (the Special Purpose Vehicle) focused on studying diffusion, submarine wakes and acoustics in the late 1950s. Over time there has since been an impressive number of systems developed with ever increasing, expanding capabilities (Yoerger et al., 1998; Davis et al., 2003; Caress et al., 2008; German et al., 2008). These systems were large and often required large dedicated engineering teams and support

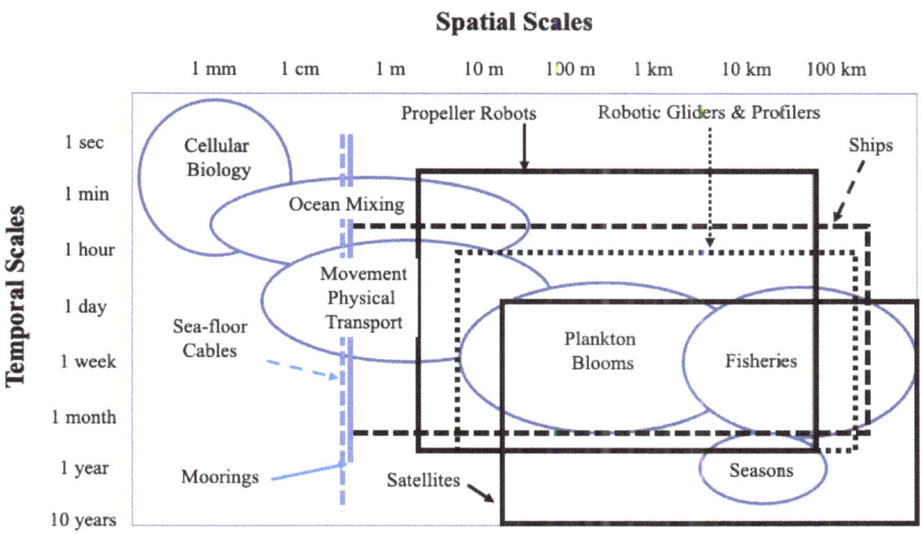

Figure 1 The time (y-axis) and horizontal surface ocean space (x-axis) scales at which critical ocean processes operate (denoted by circles). The sampling capabilities associated with specific sampling platforms is outlined by squares and lines. Modified from Schofield et al. (2014).

infrastructure. This has limited where the technology could be used. RUMFS has focused on the development of systems that could be used by a broader community not requiring a major support team.

One key technology that was tested and demonstrated at the Rutgers University Marine Field Station was propeller-driven underwater vehicles. Propeller-driven autonomous underwater vehicles (AUVs) are powered by batteries or fuel cells and can operate in water as deep as 6000 m. The AUVs relay data and mission information to shore via satellite. Between position fixes and for precise maneuvering, inertial navigation systems are often available onboard the AUV to measure the acceleration of the vehicle, and combined with Doppler velocity measurements, are used to measure the rate of travel. A pressure sensor measures the vertical position. AUVs can move against currents nominally at 3–5 knots, and can therefore systematically and synoptically survey a particular line, area, and/or volume with high precision. The system that was used extensively was the Remote Environmental Monitoring Units (REMUS), which is a low cost AUV designed by the Oceanographic Systems Lab, that was tested during a series of coastal predictive experiments (Glenn et al., 2004). As the REMUS system matured it has served a wide range of applied needs including mine detection in forward deployed regions and surveying urban aqueducts for leaks (https://www.whoi.edu/what-we-do/explore/underwater-vehicles/auvs/remus/).

Another class of mobile autonomous systems initially tested and demonstrated at the Rutgers University Marine Field Station includes gliders which are a type of autonomous underwater vehicle that use small changes in buoyancy in conjunction with wings to convert vertical motion to horizontal motion, and thereby propel itself forward with very low-power consumption (Schofield et al., 2008). Gliders follow a saw-tooth path through the water, providing data on large temporal and spatial scales. They navigate with the help of periodic surfacing for Global Positioning System (GPS) fixes, pressure sensors, tilt sensors, and magnetic compasses. Using buoyancy-based propulsion, gliders have a significant range and duration, with missions lasting up to a year and covering over 3500 km of range. Since the initial tests of the gliders at the Rutgers University Marine Field Station, Rutgers scientists have operated gliders all over the world (Figure 2). To date, these measurements represent 744 deployments representing 21,330 days at sea.

A major challenge is deploying science sensors in the ocean with sufficient power (often based on the available power provided by onboard batteries) for sustained deployments. Additionally, data transmission, even

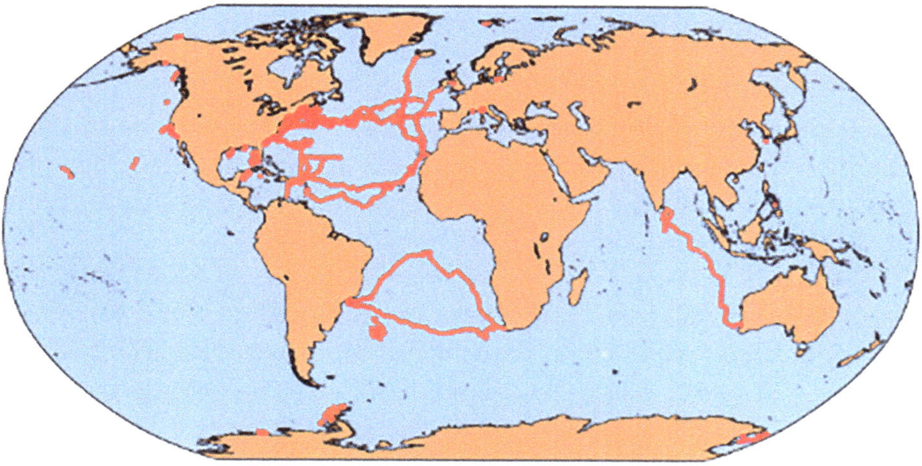

Figure 2 The global deployments of gliders in coastal and open ocean waters by Rutgers scientists. The first ocean deployments of these gliders were conducted at RUMFS.

with global cell phone capabilities, can be limited for many science platforms especially for sensors that have high data bandwidth. One potential strategy is to deploy electro-optical cables in the ocean. These cables can provide high power out to remote locations in the ocean and provide effectively unlimited data transmission back to shore. The first electro-optical dedicated sea floor cable for science was deployed at RUMFS in 1998. The system was called the Long term Ecosystem Observatory and the cable provided power for data telemetry at 15 meters depth. The system was known as LEO-15. LEO-15 system was a unique first of a kind capability in the world, and its success (Schofield et al., 2002) resulted in the deployment of numerous science dedicated seafloor cables around the world including the United States, Canada, Japan, and Europe.

A major difficulty working in the ocean is that it moves in complex patterns that are difficult to resolve using measurements at a single point in space. The need for sustained spatial data on circulation patterns is a critical need. The Rutgers University Marine Field Station has been at the forefront of developing the high frequency radar measurements capable of continuous mapping of the surface currents of the ocean. The high-frequency radar measures ocean surface current velocities over hundreds of square miles simultaneously. Each site measures the radial components of the ocean surface velocity directed towards or away from the site and the estimated velocity components allow surface currents (upper meter of water column) to be estimated. These systems are cost effective and have many applied uses. The network initially deployed at RUMFS has grown into the world's most advanced global surface current mapping system collecting data spanning from Cape Cod to Cape Hatteras for decades (Figure 3). The data allows scientists to understand the ocean circulation and provides data that has practical applications. For example, the US Coast Guard actively uses the radar data to plan search and rescue operations.

Ocean observations are complemented with ocean models that can help synthesize what the data means. Over the last 30 years, there have been significant developments in three-dimensional numerical models for the ocean. Many models exist spanning from global ocean scales down to the scale of individual estuaries. Models vary in their coordinate system (linear, spherical, and others), resolution in space and time, complexity (i.e., number of state variables), and the parameterization of key processes within the model. There are several excellent texts that outline many of the details of numerical ocean modelling and one key lesson is that the choice of particular modelling approach depends on its intended application and on the available computational resources. Although an exhaustive list of the ocean models is beyond the scope of this text, several classes are described in the following paragraphs.

Mechanistic models are simplified models used to study a specific process, and are used to provide insight into the underlying processes influencing the physics, chemistry, and biology of the ocean. These models are most often constructed as a learning tool in order to assess processes and feedbacks within marine systems.

Simulation models are complex and describe three dimensional (3-D) ocean processes using the continuity and momentum equations. For this reason they are called the primitive equation models. These models can be used to simulate many processes, including ocean circulation, mixing, waves, and responses to external forces (such as storms). All these models are constructed using different assumptions. Additionally, the resolution of the models requires that the trade-offs of the computation burdens be measured against the processes that need to be simulated. For example, if the model must resolve mesoscale eddies, it will require the resolution of a few tenths of a degree of latitude and longitude. In contrast, most primitive equation climate models have much coarser horizontal resolution as they were designed to study large-scale hydrographic structure, climate dynamics, and water-mass formation over decadal time scales. However, for a specific question, there are climate models with sufficient resolution to resolve mesoscale eddies if one is willing to accept the computation cost. Simulations are also constructed for coastal systems and can resolve coastal currents, tides, and storm surges.

Increasingly the biological and ocean chemistry models are being coupled into these 3-D simulation models. Although these biogeochemical models are rapidly improving, there is unfortunately no set of "primitive" equations yet capable of describing biological and chemical systems in the ocean; as these models evolve, however, they will be increasingly useful tools for managing living resources and water quality in the ocean.

Often models of varying resolutions are combined. Coarser-scale global or basin-scale models provide outer boundary inputs to higher resolution nested models, which allows a myriad of processes to be modelled with a lower computation burden but allow a range of processes to be simulated even if they require high resolution. This is often the case for coastal and continental shelf models. The advantage of this downscaling approach is that it allows basin scale models to resolve large-scale forcing that drives the regional to local-scale processes that are effectively modelled by a higher resolution model. The approach by which one links these models is a difficult problem and remains an area of active research.

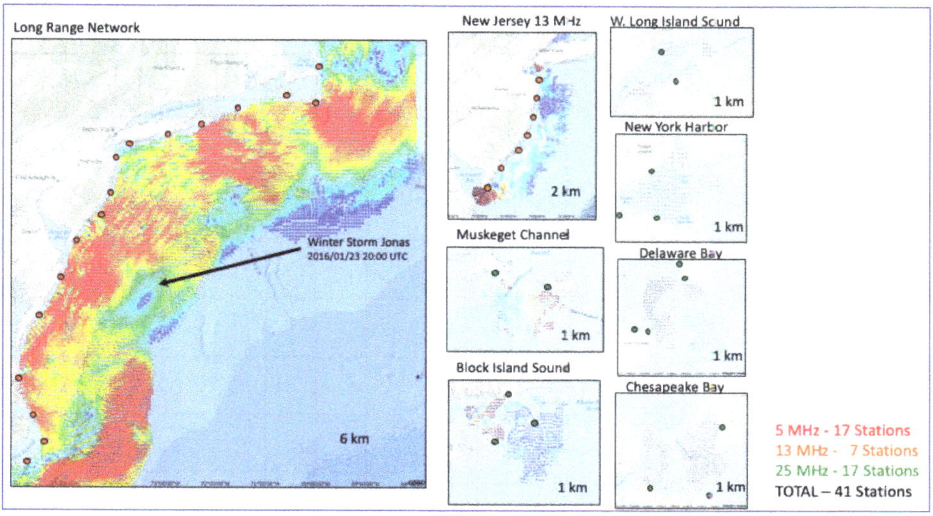

Figure 3 The high frequency radar networks spanning the northeast United States. Different radar systems can have different frequencies and this determines the resolution of the current measurement. The existing network provides hourly current updates. The resolution of the system increases closer to the coast line.

Data assimilation is an approach by which model simulations are constrained by observations. For example, model calculations and observations of temperature and salinity can be compared, and then the model can be "adjusted" based on the mismatch. This is a difficult problem as 1) it represents an inverse problem (where a finite number of observations are used to estimate a continuous field), 2) many of the ocean processes of interest are non-linear, and 3) the observations and models both have unknown errors. These approaches allow modelers to increase the forecast skill of their models by essentially keeping the models "on track" if the observations and data assimilation approaches can be provided in a timely fashion. Many in the ocean modelling community are focusing on using these approaches to increase model forecast skill as it determines how well these approaches will serve a wide range of science, commercial, and government needs.

The combination of the observation and modeling capabilities combined through the data assimilation models offer the opportunity to develop smart sampling networks. These networks are often referred to as a sensor web, where a distributed sampling network is linked through web-based services allowing for the integrated system to sample the ocean as efficiently as possible. These ocean sensor webs are an evolving tool for research, but some of the original ocean sensor webs were developed and tested at RUMFS (Glenn et al., 2004; Schofield et al., 2004; Schofield et al., 2008). One of the initial demonstrations was conducted during a series of coastal predictive skill experiments at RUMFS (Figure 4). The efforts linked forecast models with a diverse array of observing systems. These network have resulted in the global array of systems that are growing in the complexity and automation (Schofield et al., 2010). The evolving networks are becoming central to modern oceanography.

Acknowledgments

The efforts and work described reflects the generous support of the Office of Naval Research, National Oceanic Atmospheric Administration, NASA, National Science Foundation, the New Jersey Department of Environmental Protection, Rutgers University, and most of all the support of the people and government of the great state of New Jersey. We also want to thank the countless students (undergraduate and graduate) and passionate staff that have made RU COOL what it is. To those staff and students, we salute you!

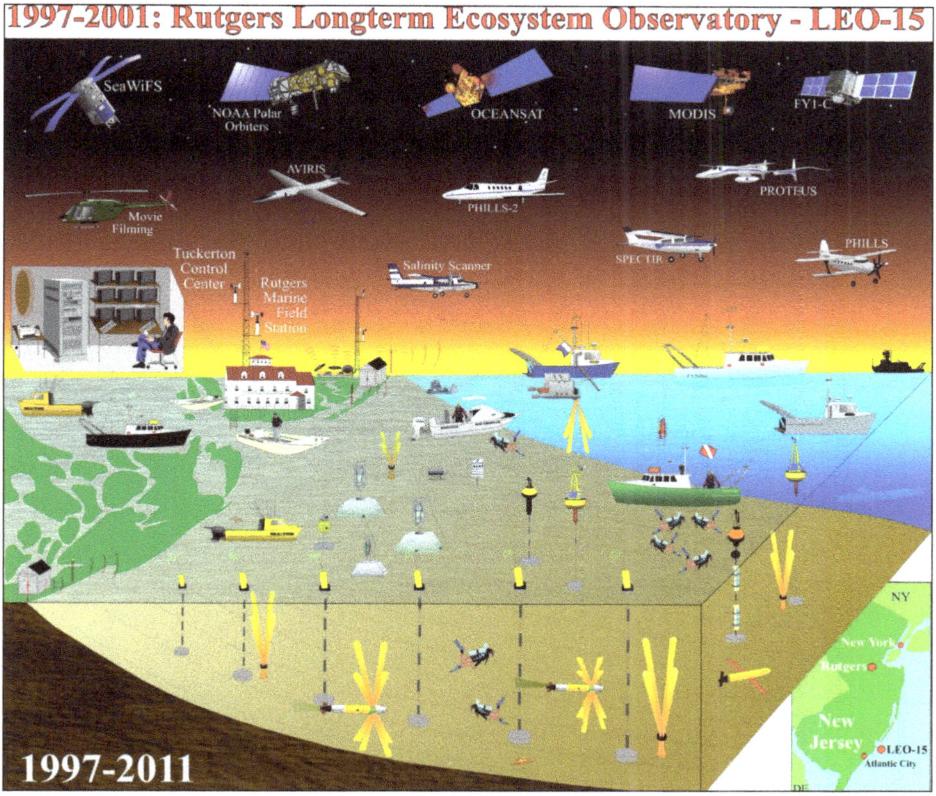

Figure 4 The sensor web developed at the Rutgers University Marine Field Station that integrated diverse technologies with numerical forecast models.

References

Able, K. W. (2015). *Station 119: From lifesaving to marine research*. Down the Shore Publishing.

Able, K. W. (2020). *Beneath the surface: Understanding nature in the Mullica Valley Estuary*. Rutgers University Press.

Caress, D. W., Thomas, H., Kirkwood, W. J., McEwen, R., Henthorn, R., Clague, D. A., Paull, C. K., Paduan, J., & Maier, K. L. (2008). High-Resolution multibeam, sidescan, and subbottom surveys using the MBARI AUV D. Allan B. Marine Habitat Mapping Technology for Alaska. J.R. Reynolds and H.G. Greene (eds.) 47 Alaska Sea Grant College Program, University of Alaska Fairbanks. doi:10.4027/mhmta.2008.04.

Davis, R. E., Eriksen, C. C., & Jones, C. P. (2003). Autonomous buoyancy-driven underwater gliders. In G. Griffiths (Ed.), *Technology and applications of autonomous underwater vehicles* (37–58). Taylor and Francis.

Dickey, T., Fyre, D., McNeil, J., Manov, D., Nelson, N., Sigurdson, D., Jannasch, H., Siegel, D., Michaels, T., & Johnson, R. (1998). Upper-ocean temperature response to hurricane Felix as measured by the Bermuda testbed mooring. *American Meteorological Society, 126,* 1195–1201.

German, C. R., Yoerger, D. R., Jakuba, M., Shank, T. M., Langmuir, C. H., & Nakamura, K. (2008). Hydrothermal exploration with the Autonomous Benthic Explorer. *Deep Sea Research, 55*, 203–219.

Glenn, S. M., Arnone, R., Bergmann, T., Bissett, W. P., Crowley, M., Cullen, J., Gryzmski, J., Haidvogel. D., Kohut, J., Moline, M. A., Oliver, M., Orrico, C., Sherrell, R., Song, T., Weidemann, A., Chant, R., & Schofield, O. (2004). The biogeochemical impact of summertime coastal upwelling in the Mid-Atlantic Bight. *Journal of Geophysical Research, 109.* C12S02, DOI:10.1029/2003JC002265.

Glenn, S. M., & Schofield, O. (2003). Observing the oceans from the COOL-room: Our history, experience, and opinions. *Oceanography, 16*(4), 37–52.

Glenn, S., Schofield, O., Dickey, T. D., Chant, R. Kohut, Barrier, H., Bosch, J., Bowers, L., Creed, E., Haldeman, C., Hunter, E., Kerfoot, J., Mudgal, C., Oliver, M., Roarty, H., Romana, E., Crowley, M., Barrick D., & Jones C. (2004). The expanding role of ocean color and optics in the changing field of operational oceanography. *Oceanography 17*, 86–95.

Glover, H. E., Prezelin, B. B., Campbell, L., Wyman, M., & Garside, C. (1998). A nitrate-dependent Synechoccus bloom in surface Sargasso Sea water. *Nature, 331*, 161–64

National Research Council. (2011). *Critical infrastructure for ocean research and societal needs in 2030*. National Academy Press.

RU COOL. The Rutgers University Center for Ocean Observing Leadership.

Schofield, O., Arnone, R., Bissett, W. P., Dickey, T., Davis, C., Finkel, Z., Oliver, M., & Moline, M. A. (2004). Watercolors in the coastal zone: What can we see? *Oceanography 17*, 28–37.

Schofield, O., Bergmann, T., Bissett, W. P., Grassle, F., Haidvogel, D., Kohut, J., Moline, M., & Glenn, S. (2002). Linking regional coastal observatories to provide the foundation for a national ocean observation network. *Journal of Oceanic Engineering, 27*(2), 146–54.

Schofield, O., Chant, R., Cahill, B., Castelao, R., Gong, D., Kahl, A., Kohut, J., Montes-Hugo, M., Ramadurai, R., Ramey, P., Xu, Y., & Glenn, S. M.

(2008). Seasonal forcing of primary productivity on broad continental shelves. *Oceanography, 21*(4): 104–117.

Schofield, O., Glenn, S., Orcutt, J., Arrott, M., Brown, W., Signell, R., Moline, M. A., Chao, Y., Chien, S., Thompson, D., Balasuriya, A., & Oliver, M. (2010). Automated sensor networks to advance ocean science. *Transactions of the American Geophysical Union, 91*(39): 345–46. doi:10.1029/2010EO390001.

Schofield, O., Kohut, J., Saba, G., Xu, Y., Wilkin, J., & Glenn, S. M. (2014). Ocean observing and prediction. *Encyclopedia of Natural Resources.* Wang, Y. Q. (editor). Taylor Francis. doi:10.1018/E-ENRW-120048087 802–807.

Schofield, O., & Tivey., M. (2004). Building a window to the sea: Ocean Research Interactive Observing Networks (ORION). *Oceanography 17*, 105–111.

Yoerger, D. R., A. M. Bradley, B. B. Walden, H. Singh, & Bachmayer, R. (1998). Surveying a subsea lava flow using the Autonomous Benthic Explorer (ABE). *International Journal of Systems Science, 29*(4), 1031–44.

Acknowledgments

I have been gathering ideas, data, images and recollections for this book over several decades. These years have been filled with collaborative research, observations and analysis along Seven Bridges Road. This could only have occurred with the assistance of numerous individuals, organizations, and the general public. Most notably these include the exceptional contributions of individuals from the Rutgers University Marine Field Station which at last count included over 20 postdoctoral fellows, over 40 graduate students, more than 120 technicians and dozens of summer interns over the period from the late 1980s to the present. Of these, my faculty colleagues have been especially helpful including "Motz" Grothues over the last two decades, who now has become the Director. Others from the Station who have made major contributions over time to this effort include Roland Hagan and Carol Van Pelt. In recent years, Oscar Schofield has managed the Station.

Another group deserving recognition are the numerous volunteers (over 60 between 2012–2023) from the Station and the Jacques Cousteau National Estuarine Research Reserve. These have included students from a variety of institutions including Rutgers University, Stockton University, Montclair State University, Rider University, Stevens Institute, Boston College, West Virginia University and Michigan University. Several volunteers transcribed tapes from field trips (notably Rose Faiss) or participated in field sampling and sample sorting, sometimes at all hours of the day and night (including Pat Filardi, Tom Siciliano, Steve Zeck and others too numerous to mention). Many of these activities have been organized by Roland Hagan. Other groups assisted in lending an historical perspective for the area including the Tuckerton Historical Society and the Tuckerton Seaport, particularly Paul Hart and Jim Allen. Pete McLane allowed access to his correspondence regarding his leadership in the history of the development of the Great Bay Wildlife Management Area. Exceptional contributions came from helicopter flights (provided by Judy Redlawsk in her own helicopter) and drone images (contributed from Rob Auermuller and Thomas Johnson with their drones). Their excellent images are obvious throughout the chapters and provide a broader context to observations on the ground and on the water (kayaking with Pat Filardi). The images of others along the road were by other authors (Bob Birdsall and Terry O'Leary) who graciously provided access to their images from their own trips along Seven Bridges Road. Other images came from a variety of sources and these are indicated in the individual figure captions; the exceptions are those by the author. Another group that broadened the scope and quality of this book were the authors of the appendices (Andrew and John Stanzeski, Oscar Schofield and Scott Glenn, Steve Spahn).

Numerous other individuals provided answers to specific, diverse questions in a timely manner that helped make this treatment more complete (Mike Mangum, Bob Keeney, Kathy Conrad, Alan Mounier, Bari Page, Anna Serbek, David Robinson, Pete Stemmer, Ryan Sherwood, Paul Hart, Jim Allen, John Schriever, Scott Sanford, Jane Uptegrove, Peter Zmek, Steve Dodson, Ron Marr, Kaitlin Kennedy, Dan Melega, Fernando Fuentes, Norb Psuty, Joe Arsenault, Matt Matusky). Others helped with collection of historical artifacts and data (John Yates, Maggie Shaw, Miranda Rosen). Over the years, numerous scientists have visited from throughout New Jersey but also from Pennsylvania, West Virginia, Michigan, Maryland, Virginia, North Carolina, South Carolina, Florida, Alaska, Louisiana, and Canada.

In the critical stages of the book preparation several individuals provided much needed assistance including organization of numerous images from a variety of sources (Tara Collins), preparation of maps and other illustrations (Ryan Larum), word processing (Christine Moran, Betty Lesher), and hunting down references and data files (Miranda Rosen, Maggie Shaw). Patsy and Larry Schmalz and Alice and Jerry Gob provided quiet places to write.

The editors (Tom Kinsella, Paul W. Schopp, and their student interns) at the South Jersey Culture & History Center assisted with the organization and publication of the final product. Tom and Paul were particularly helpful in the final stages of this book.

References by Chapter

Chapter 1 References

Able, K. W. (2016). Natural history: An approach whose time has come, passed, and needs to be resurrected. *ICES Journal of Marine Science, 73*(9), 2150–55.

Able, K. W. (2020). *Beneath the surface: Understanding nature in the Mullica Valley Estuary*. Rutgers University Press.

Bartholomew, G. A. (1986). The role of natural history in contemporary biology: Natural history can help pose questions for and supply integrative lines between different biological disciplines. *BioScience, 36*(5), 324–29.

Billick, I., & Price, M. V. (2010). *The ecology of place: Contributions of place-based research to ecological understanding*. The University of Chicago Press.

Greene, H. W. (2005). Organisms in nature as a central focus for biology. *Trends in Ecology and Evolution, 20*(1):23-27.

National Research Council of the National Academies. (2014). *Enhancing the value and sustainability of field stations and marine laboratories in the 21st century*. The National Academies Press.

Chapter 2 References

Blackman, L. (1963). *History of Little Egg Harbor Township: From the first settlement to the present time*. Trenton Printing Company, Inc.

Cook, G. H. (1868). *Geology of New Jersey*. New Jersey Geological Survey.

Cross, D. (1941). *The archaeology of New Jersey, Vol. 1*. Archaeological Society of New Jersey and New Jersey State Museum.

Daddario, J. J. (1961). A lagoon deposit profile near Atlantic City, New Jersey. *Bulletin of the New Jersey Academy of Science, 6*, 7–14.

Fleming, G. D. (2005). *Brotherton: New Jersey's first and only Indian reservation and the communities of Shamong and Tabernacle that followed*. Plexus Publishing, Inc.

French, H. M., & Demitroff, M. (2001). Cold-climate origin of the enclosed depressions and wetlands ("Spungs") of the Pine Barrens, southern New Jersey, USA. *Permafrost and Periglacial Processes, 12*, 337–50.

Goos, N. R. (2011). A very large British military investment for very little practical profit. *Atlantic County Historical Society and Genealogical Journal,*

16(4), 58–68.

Guo, Z., & Psuty, N. P. (1997). Flood-tide deltaic wetlands: Detection of the sequential spatial evolution. *Photogrammetric Engineering and Remote Sensing, 63*(3), 273–80.

Harper, D. P. (2013). *Roadside geology of New Jersey*. Mountain Press Publishing Company.

Jordan Jr., F. (1906). Fishing stations on the New Jersey coast—pile dwellings. In *Aboriginal fishing stations on the coast of the Mid-Atlantic states* (pp. 10–18). New Era Printing Company.

Katz, J. W. (1954, February 26). Forest 40 feet under ground. *Newark Evening News*.

Kemp, A. C., Horton, B. P., Vane, C. H., Bernhardt, C. E., Corbert, D. R., Engelhart, S. E., Anisfeld, S. C., Parnell, A. C., & Cahill, N. (2013). Sea-level change during the last 2500 years in New Jersey, USA. *Quaternary Science Reviews, 81*, 90–104.

Lurie, M. N., & Veit, R. F. (2016). *Envisioning New Jersey: An illustrated history of the Garden State*. New Jersey Council for the Humanities.

Mazzagetti, D. (2018). *The Jersey Shore: The past, present and future of a national treasure*. Rutgers University Press.

Miller, K. G., Sugarman, P. J., Browning, J. V. et al. (2009). Sea-level rise in New Jersey over the past 5000 years: Implication to anthropogenic changes. *Global and Planetary Change, 66*, 10–18.

Miller, K. G., Kopp, R. E., Horton, B. P., Browning, J.V., & Kemp. (2013). A geological perspective on sea-level rise and its impacts along the U.S. mid-Atlantic coast. *Earth's Future, 1*, 3–18.

Mounier, R. A. (1999). A brief history of investigation of the Tuckerton Shell Mound. *Bulletin of the Archaeological Society of New Jersey, 54*, 43–44.

Mounier, R. A. (2002). *Looking beneath the surface: The story of archaeology in New Jersey*. Rutgers University Press.

Mounier, R. A. (2004). The Tuckerton Shell Mound: A lasting fertilizer. In *From the Arctic to Avalon: Transforming the history of Northeastern North America*. Presentation at the Conference in St. John's Newfoundland.

Mounier, R. A., & Arsenault, J. R. (2019). Discovery and dating of a cucurbit seed from the Tuckerton Shell Mound, Little Egg Harbor Township, Ocean County, N.J. *Bulletin of the Archaeological Society of New Jersey, 74*, 97–102.

Plusquellec, P. L. (1966). *Coastal morphology and changes of an area between Brigantine and Beach Haven Heights, New Jersey* [Master's thesis, University of Illinois].

Psuty, N. P. (1986). Holocene sea level in New Jersey. *Physical Geography, 7*(2),

156–67.
Racine and Laramie Tobacconist. http://www.racineandlaramie.com/.
Skinner, A., & Schrabisch, M. (1913). A preliminary report of the archaeological survey of the State of New Jersey. *Geological Survey of New Jersey Bulletin 9*. Trenton, New Jersey.
Stanzeski, A. J. (1996). Two decades of radiocarbon dating from the New Jersey shore. Bulletin of the *Archaeological Society of New Jersey, 51*, 42–45.
Stanzeski, A. J. (2001). The Tuckerton Shell Mound. *Bulletin of the Archaeological Society of New Jersey, 56*, 47–50.
Stanzeski, A. J. (in review). The builders of the Tuckerton Shell Mound 2000 -/+ years ago: Archaeology: Pennella Site. *Bulletin of the Archaeological Society of New Jersey*.
Stanzeski, A., & Stanzeski, J. (Appendix 1). Coastal woodland sites in Mystic Island, New Jersey. (Osborn Island, Graveling Point and Wells Island areas).
Uptegrove, J., Monteverde, D. H., Whitesell, L. A., Stanford, S. D., & Waldner, J. J. (2015). High-resolution seismic tracing of the transgression across the coastal interface, Brigantine, New Jersey. *Geophysics, 80*(3), WB43–WB63.
Veit, R. (2012). Ch. 1: Setting the stage: archaeology and the Delaware Indians, a 12,000 year odyssey. In M. N. Lurie and R. F. Veit (Eds.), *A history of the Garden State* (pp. 6–32). Rutgers University Press.
Williams, L. E. (2010). Caught in the middle: New Jersey's Indians and the American Revolution. In M. N. Lurie (Ed.), *A New Jersey anthology* (pp. 109–119), Rivergate Books. Rutgers University Press.

Chapter 3 References

Able, K. W. (2017). Mapping the Mullica Valley: Natural history landscapes. *SoJourn, 2*(1), 33–44.
Able, K. W. (2017). Manufacturing from menhaden: A history of the Mullica Valley. *SoJourn, 1*(2), 75–82.
Able, K. W. (2020). History and ecology of salt marsh ditches in the Mullica Valley. *SoJourn, 5*(1), 7–17.
Able, K. W. (2020). *Beneath the surface: Understanding nature in the Mullica Valley Estuary*. Rutgers University Press.
Anonymous. (2018). South Jersey's revolutionary battles, skirmishes, and future research. *SoJourn*, Summer 2018, 127–36.
Blackman, L. (1963). *History of Little Egg Harbor Township, Burlington County, N.J. From its first settlement to the present time*. Trenton Printing Company, Inc.

Braddock-Rogers, K. (1938). Saltworks of New Jersey during the American Revolution. *Journal of Chemical Education, 15*, 586–92.

Candeletti, R., & Candelitti, T. M. (1990). A cost comparison between Open Marsh Water Management and chemical larviciding utilizing modern mosquito control equipment. *Proceedings of the Seventy-Seventh Annual Meeting of the New Jersey Mosquito Control Association*, 53–57.

Dodson, S. (2018). *Tuckerton: A newspaper history, 1852–1917*. Tuckerton Historical Society.

Epstein, C. M. (2018). When Mad Anthony came to South Jersey: Civilian experience during the American Revolution. *SoJourn*, Summer 2018, 81–94.

Ferrigno, F., & Jobbins, D. M. (1968). Open Marsh Water Management. *Proceedings of the Fifty-Fifth Annual Meeting of the New Jersey Mosquito Control Association*, 104–115.

Ford, S. E. (1947). History and present status of molluscan shellfisheries from Barnegat Bay to Delaware Bay. *NOAA Technical Report NMFS, 127*, 119–40.

Goode, G. B. (1884). The fisheries and fishery industries of the United States. *U.S. Commercial Fisheries Report*, Section 1, Part III.

Goos, N. R. (2011). A very large British military investment for very little practical profit. *Atlantic County Historical Society and Genealogical Journal, 16*(4), 58–68.

Goos, N. R. (2018). Born a peacemaker, became a patriot: 1st Lieutenant Jeremiah Leeds, New Jersey's 3rd Battalion, Gloucester County militia during the Revolutionary War, with genealogical notes on the Leeds Family. *SoJourn*, *3*(1), 107–119.

Harness, J. A. (2018). Knight at Egg Harbor. *SoJourn*, *3*(1), 31–40.

Harness, J. A. (2018). The Lord's orders. *SoJourn*, *3*(1), 95–106.

Heritage Studies Inc. (1985). *Archaeological data recovery at the Willits/Andrews Farmstead, Little Egg Harbor Township, Ocean County, New Jersey*. Trenton, New Jersey.

Kemp, F. W. (1993). *A nest of rebel pirates* (second edition). The Laureate Press.

Lloyd, J. B. (1990). *Six miles at sea: A pictorial history of Long Beach Island New Jersey*. Down the Shore Publishing/The Sandpaper.

Lloyd, J. B. (1994). *Eighteen miles of history on Long Beach Island*. Down the Shore Publishing/The Sandpaper.

Lloyd, J. B. (2005). *Two centuries of history on Long Beach Island*. Down the Shore Publishing, Harvey Cedars, New Jersey.

Mazzagetti, D. (2018). *The Jersey Shore: The past, present and future of a national treasure*. Rutgers University Press.

McCay, B. (1980). A footnote to the history of New Jersey fisheries: Menhaden as food and fertilizer. *Proceedings of the New Jersey Historical Society, 98*(3/4), 212–22.
Miller, P. S. (2000). *Ocean County: Four centuries in the making*. Ocean County Cultural and Heritage Commission.
Moonsammy, R. Z., Cohen, S. S., & Williams, L. E. (1987). *Pinelands folklife*. Rutgers University Press.
Nagiewicz, S. D. (2016). *Hidden history of maritime New Jersey*. The History Press, Charleston, South Carolina.
Nunn, R. M. (1967). *The Goldschmidt Wireless of Tuckerton, New Jersey* [Bachelor of Arts thesis, Albright College].
Pierce, A. D. (1960). *Smugglers' woods: Jaunts and journeys in colonial and revolutionary New Jersey*. Rutgers University Press.
Stephenson Jr., E. T., & Ueckermann, G. W. (2014). A history of the Pleasantville Meadows. Part III: 1970 to present. *The Atlantic County Historical Society, Sixty-Seventh Yearbook*, 41–53.
Stryker, W. S. (1894). *The affair at Egg Harbor, New Jersey, October 15, 1778*. Naar, Day and Naar.
Uekermann, G. W., & Stephenson Jr., E.T. (2012). A history of the Pleasantville Meadows Part I: Prehisitory to 1850. *The Atlantic County Historical Society, Sixty-Fifth Yearbook*, 39–56.
Weiss, H. B., Kemble, H. R., & Carre, M. T. (1974). *Whaling in New Jersey*. New Jersey Agricultural Society.
Weiss, H. B., & Weiss, G. M. (1959). *The revolutionary saltworks of the New Jersey Coast*. The Past Times Press.
Wrobleski, J. E. (2017). The affair at Egg Harbor: Massacre of the Pulaski Legion. *Journal of the American Revolution*, https://althingsliberty.com.
Yates, J. H. (2024). Hatfield's and Willit's stores/hotels on Sheepshead Meadows: Dancing with Leah Blackman's sister-in-law. *SoJourn, 8*, 63–87.

Chapter 4 References

Able, K. W. (2015). *Station 119: From lifesaving to marine research*. Down the Shore Publishing.
Able, K. W. (2020). *Beneath the surface: Understanding nature in the Mullica Valley Estuary*. Rutgers University Press.
Collette, B. B., & Klein-MacPhee, G. (2002). *Fishes of the Gulf of Maine* (third edition). Smithsonian Institution Press.
Dodson, S. (2019). *The Tuckerton Beacon, 1917–1929: World War I, through the Roaring Twenties to the Great Depression*. Tuckerton Historical Society.

Dodson, S. (2021). *The Tuckerton Beacon, 1929–1941: Part Two: The Great Depression to World War II*. Tuckerton Historical Society.

Hildebrand, S. F., & Schroeder, W. E. (1928). Fishes of Chesapeake Bay. In *Bulletin of the United States Bureau of Fisheries*, 43, Part I.

Lawrence, S. (1984). *The Audubon Society field guide to the natural places of the Mid-Atlantic States: Coastal*. Pantheon Book.

Nature Conservancy of New Jersey. (1998). The Nature Conservancy's guide to preserves and projects in New Jersey. Scott Printing, New Providence, New Jersey.

New Jersey Intracoastal Waterway. (1938). *Survey of 1936–1937—Little Egg Harbor to Great Bay*. U. S. Engineer Office, Philadelphia, PA.

Savadove, L., Buchholz, M. T., & Mazzella, S. (2019). *Great Storms of the Jersey Shore* (second edition). Down The Shore Publishing.

Waldman, J. (2013). *Heartbeats in the muck: The history, sea life, and environment of New York Harbor*. Empire State Edition. Fordham University Press.

Chapter 5 References

Able, K. W. (2015). *Station 119: From lifesaving to marine research*. Down the Shore Publishing.

Able, K. W. (2017). Manufacturing from menhaden: A history of the Mullica Valley. *SoJourn 1*(2), 75–82.

Able, K. W. (2020). *Beneath the surface: Understanding nature in the Mullica Valley Estuary*. Rutgers University Press.

Able, K. W. (2020/2021). The unnatural history of South Jersey lagoons. *SoJourn 5*(2), 49–54.

Davis, M. M. (1999). 1955: The Parkway and the Shore. In M. T. Buchholz. *Shore Chronicles: Diaries and travelers' tales from the Jersey Shore* (pp. 356–60). Down the Shore Publishing.

Dodson, S. (2021). *The Tuckerton Beacon, 1929–1941: Part One—The Great Depression to WWII*. Tuckerton Historical Society.

Dodson, S. (2022). *Tuckerton, LBI and Southern Ocean County 1945–1954. Part One. Post WWII to the Garden State Parkway*. Tuckerton Historical Society.

Easy, T. (2001). *Shorebird shooting on Sheepshead Flats*. Ocean County Decoy and Gunning Show Program.

Garrity-Blake, B. J. (1994). *The fish factory: Work and meaning for black and white fishermen of the American menhaden industry*. The University of Tennessee Press.

Gaul, G. M. (2019). *The geography of risk: Epic storms, rising seas, and the cost of America's coast*. Sarah Crichton Books.

Lloyd, J. B. (1994). *Golden age of gunning.* Old Time Barnegat Bay Decoy and Gunning Show Program.
Lloyd, J. B. (1990). *Six miles at sea: A pictorial history of Long Beach Island New Jersey.* Down the Shore Publishing and the Sandpaper, Inc.
Lloyd, J. B. (1994). *Eighteen miles of history on Long Beach Island.* Down the Shore Publishing and the Sandpaper, Inc.
Lloyd, J. B. (2005). *Two centuries of history on Long Beach Island.* Down the Shore Publishing and the Sandpaper, Inc.
McKenzie, C. L. (1992). *The fisheries of Raritan Bay.* Rutgers University Press.
Miller, P. S. (2000). *Ocean County: Four centuries in the making.* Ocean County Cultural and Heritage Commission.
Publications Committee of the Ocean County Historical Society. (1997). *Downshore from Manahawkin to New Gretna.* Arcadia Publishing.
Schneider, A., Lopez-Duarte, P. C., & Able, K. W. (2025). Long-term patterns of juvenile blue crab (*Callinectus sapidus*) recruitment in a small, mid-Atlantic estuary, U.S.A. *Journal of Crustacean Biology, 45*(2): ruaf024.

Chapter 6 References

Able, K. W. (1999). Measures of juvenile fish habitat quality: Examples from a National Estuarine Research Reserve. In L. R. Benaka (Ed.), *Fish habitat: Essential fish habitat and rehabilitation* (pp. 134–47). American Fisheries Society Symposium 22, Bethesda, Maryland.
Able, K. W., Grothues, T. M., Rackovan, J. L., & Buderman, F. E. (2014). Application of mobile dual-frequency identification sonar (DIDSON) to fish estuarine habits. *Northeast Naturalist, 21*(2), 192–209.
Able, K. W. (2015). *Station 119: From lifesaving to marine research.* Down The Shore Publishing.
Able, K. W. (2020). *Beneath the surface: Understanding nature in the Mullica Valley Estuary.* Rutgers University Press.
Able, K. W., Smith, K. J., & Hagan, S. M. (2005). Fish composition and abundance in New Jersey marsh pools: Sampling technique effects. *Northeastern Naturalist, 12*(4), 485–502.
Able, K. W., & Grothues, T. M. (2007). Diversity of estuarine movements of striped bass (*Morone saxatilis*): A synoptic examination of an estuarine system in southern New Jersey. *Fisheries Bulletin, 105*, 426–35.
Chant, R. J., Curran, M. C., Able, K. W., & Glenn, S. M. (2000). Delivery of winter flounder (*Pseudopleuronectes americanus*) larvae to settlement habitats in coves near tidal inlets. *Estuarine, Coastal and Shelf Science, 51*, 529–41.

Charlesworth Jr., L. J. (1968). *Bay, inlet, and nearshore marine sedimentation: Beach Haven–Little Egg Inlet region, New Jersey* [Doctoral dissertation, The University of Michigan].

Gao, Y., Kennish, M. J., & Flynn, A. M. (2007). Atmospheric nitrogen deposition to the New Jersey coastal waters and its implications. *Ecological Applications, 17*(5), 531–41.

Glenn, S. M., Boicourt, W., Parker, B., & Dickey, T. D. (2000). Operational observation networks for ports, a large estuary and an open shelf. *Oceanography, 13*(1), 12–23.

Grassle, J. F., Glenn, S., & Von Alt, C. (1998). Ocean observing systems for marine habitats. *Proceedings, Ocean Community Conference 1998*, Marine Technology Society, November, 567–70.

Grothues, T. M., Able, K. W., McDonnell, J., & Sisak, M. M. (2005). An estuarine observatory for real-time telemetry of migrant macrofauna: Design, performance and constraints. *Limnology and Oceanography Methods, 3*, 275–89.

Grothues, T. M., & Able, K. W. (2007). Scaling acoustic telemetry of bluefish in an estuarine observatory: Detection and habitat use patterns. *Transactions of the American Fisheries Society, 136*, 1511–19.

Lovy, J., Lewis, N. L., Friend, S. E., Able, K. W., et al. (2020). Host, seasonal and habitat influences on incidence of *Lernaeenicus radiatus* (Copepoda: Pennellidae) in the mid-Atlantic Bight. *Marine Ecology Progress Series, 642*, 83–101.

Schofield, O., Bergmann, T., Bissett, P., Grassle, J. F., et al. (2002). The long-term ecosystem observatory: An integrated coastal observatory. *IEEE Journal of Oceanic Engineering, 27*(2), 146–54.

Strakosch, K. A. (1992). *The effects of soil water movement on the productivity and elemental composition of a New Jersey salt marsh population of short form* Spartina alterniflora [Master's thesis, Rutgers University].

Von Alt, C. J., & Grassle, J. F. (1992). LEO-15 An unmanned long term environmental observatory. *OCEANS 92 Proceedings@m_Mastering the Oceans Through Technology*, 849–54.

Walker, J. S., Khan, N. S., Shaw, T. A., Barber, D. C., Switzer, A. D., & Horton, B. P. (2023). Spatial and temporal distributions of live salt-marsh foraminifera in southern New Jersey: Implications for sea-level studies. *Journal of Foraminferal Research, 53*(1), 3–19.

Chapter 7 References

Able, K. W., Smith, J. M., & Caridad, J. (2015). American eel supply to an estuary and tributaries: Spatial variation in Barnegat Bay. NJ. *Northeastern Naturalist, 22*(1), 53–68.

Able, K. W. (2016). Skinny water: An important estuarine habitat. *Atlantic States Marine Fisheries Commission.* Habitat Hotlines Annual Issue.

Able, K. W., Valenti, J. L., & Grothues, T. M. (2017). Fish larval supply to and within a lagoonal estuary: Multiple sources for Barnegat Bay, New Jersey. *Environmental Biology of Fishes ,100*(6), 663–68.

Able, K. W., Larum, R., Hagan, R., & Johnson, T. (2018). Enhancing marsh resilience: Multiple perspectives from the Great Bay Wildlife Management Area. *Rutgers University Marine Field Station Report.*

Able, K. W. (2020). History and ecology of salt marsh ditches in the Mullica Valley. *SoJourn, 5*(1), 7–17.

Able, K. W. (2020). *Beneath the surface: Understanding nature in the Mullica Valley Estuary.* Rutgers University Press.

Able, K. W. (Winter 2020/21). *The unnatural history of South Jersey lagoons. SoJourn, 5*(2), 49–54.

Able, K. W., & Coia, G. (2017). The recovery of New Jersey's Mullica Valley. *Underwater Naturalist, 32*(2), 36–43.

Able, K. W., Dobarro, J., & Muzeni-Corino, A. M. (2010). An evaluation of boat basin dredging effects: Response of fishes and crabs in a New Jersey estuary. *North American Journal of Fisheries Management*, 30, 1001–1015.

Able, K. W., & Fahay, M. P. (2010). Chapter 7—Habitat use. In Able, K. W. and M. P. Fahay. *Ecology of estuarine fishes: Temperate waters of the Western North Atlantic.* Johns Hopkins University Press.

Able, K. W., Hagan, S. M., & Brown, S. A. (2006). Habitat use, movement and growth of young-of-the-year Fundulus spp. in southern New Jersey salt marshes: Comparisons based on tag/recapture. *Journal of Experimental Marine Biology and Ecology,* 335, 177–87.

Able, K. W., Smith K. J., & Hagan, S. M. (2005). Fish composition and abundance in New Jersey salt marsh pools: Sampling technique effects. *Northeastern Naturalist, 12*(4), 485–502.

Able, K. W., Welsh, C. J., & Larum, R. (2018). Salt marsh peat dispersal: Habitat for fishes, decapods crustaceans, and bivalves. In *Peat* (pp. 9–28). Intech Ltd., Rijeka, Croatia.

Andrews, K. M., Nanjappa, P., & Riley, S. P. D. (2015). *Roads and ecological infrastructure: Concepts and applications for small animals.* Johns Hopkins University Press.

Angelini, C., Van Montfrans, S. G., Hensel, M. J. S., He, Q., & Stillman, B. R. (2018). The importance of an underestimated grazer under climate change:

How crab density, consumer competition and physical stress can affect salt marsh resilience. *Oecologia, 187*, 205–17.

Barbier, E. B., Hacker, S. D., Kennedy, C., Koch, E. W., Stier, A. C., & Silliman, B. R. (2011). The value of estuarine and coastal ecosystem services. *Ecological Monographs, 81*(2), 169–93.

Bertness, M. D. (1999. *The ecology of Atlantic Shores*. Sinauer Associates, Inc.

Burger, J. (1996). *A naturalist along the Jersey Shore*. Rutgers University Press.

Crichton, O. W. (1960.) Marsh crab: Intertidal tunnel-maker and grass-eater. *Estuarine Bulletin, 5*, 3–10.

Cook, G. H., Smock, J. C., & Vermeule, C. C. (1888). *A topographical map of Egg Harbor and vicinity including the Atlantic shore from Barnegat to Great Egg Harbor*. Geological Survey of New Jersey, Atlas Sheet No. 16, Egg Harbor. Julius Bien and Company, Lithographers.

Coverdale, T. C., Altieri, A. H., & Bertness, M. D. (2012). Below ground herbivory increases vulnerability of New England salt marshes to die-off. *Ecology, 93*(9), 2085–94.

Day Jr., J. W., Crump, B. C., Kemp, W. M., et al. (2013). *Estuarine Ecology* (second edition). Wiley-Blackwell.

Denno, R. F., Gratton, C., Peterson, M.A., Langellotto, G. A., Finke, D. L., & Huberty, A. F. (2002). Bottom-up forces mediate natural-enemy impact in a phytophagous community. *Ecology, 83*, 1443–58.

Donnelly, J. P., Butler, J., Roll, S., Wengren, M., & Webb III, T. (2004). A backbarrier overwash record of intense storms from Brigantine, New Jersey. *Marine Geology, 210*, 107–21.

Forman, B. T. T., Sperling, D., & Bissonette, J. A. (2003). *Road ecology: Science and solutions*. Island Press.

Gaul, G. (2019). *The geography of risk: Epic storms, rising seas, and the cost of America's coasts*. Sarah Crichton Books, Farrar, Staus, and Giroux.

Guo, Q. and N. P. Psuty. (1997). Flood-tide deltaic wetlands: Detection of their sequential spatial evolution. *Photogrammetric Engineering and Remote Sensing, 63*(3), 273–80.

Holdredge, C., Bertness, M. D., & Altieri, A. H. (2008.) Role of crab herbivory in die-off of New England salt marshes. *Conservation Biology, 23*(3), 672–79.

Kennish, M. J., Bricker, S. B., Dennison, W. C., et al. (2007). Barnegat Bay–Little Egg Harbor estuary: Case study of a highly eutrophic coastal bay system. *Ecological Applications, 17*(5), S3–S16.

Lathrop, R. G., Cole, M. P., & Showalter, R. D. (2000). Quantifying the habitat structure and spatial pattern of New Jersey (U.S.A.) salt marshes under different habitat regimes. *Wetlands Ecology and Management, 8*, 163–72.

Mariotti, G., Spivak, A. C., Luk, S. Y., Ceccherini, G., Tyrrell, M., & Eagle Gonneea, M. (2020). Modeling the spatial dynamics of marsh ponds in New England salt marshes. *Geomorphology, 365*, 107262.

Meyerson, L. A., Saltonstall, K., & Chambers, R. M. (2009). *Phragmites australis* in Eastern North American: A historical and ecological perspective. In B. R. Silliman, E. D. Grosholz and M. D. Bertness, eds., *Human impacts on salt marshes: A global perspective* (pp. 57–82). University of California Press.

Moffott and Nichol Engineers. (2000). *Barnegat Bay feasibility study: Fish ladder at Lake Pohatcong and Manahawken Lake.* Army Corps of Engineers. Contract No. DACW 61-99-0-0001.

Montgomery, J. D., & Newcomb, M. R. (1975). A quantitative study of the vegetation near the Great Bay Boulevard, Tuckerton, New Jersey, vol. 3. *Ecological studies in the bays and other waterways near Little Egg Inlet and in the ocean in the vicinity of the proposed site for the Atlantic Generating Station, New Jersey. Progress report for the period January–December 1974.* Ichthyological Associates, Inc.

Patterson, G. (2009). *The mosquito crusades: A history of the American anti-mosquito movement from the Reed Commission to the first Earth Day.* Rutgers University Press.

Peterson, C. H., Able, K. W., DeJong, C. F., Piehler, M. F., Simenstad, C. A., & Zedler, J. B. (2008). Practical proxies for tidal marsh ecosystem services: Application to injury and restoration. In D. W. Sims (Ed.), *Advances in marine biology* (pp. 221–66), vol. 54. Academic Press, Elsevier Inc.

Pettengill, T. M., Crotty, S. M., Angelini, C., & Bertness, M. D. (2018). A natural history model of New England salt marsh die-off. *Oecologia, 186*, 621–32.

Rippel, T. M., Mooring, E. Q., Tomasula, J., & Wimp, G. M. (2020). Habitat edge effects decrease litter accumulation and increase litter decomposition in coastal salt marshes. *Landscape Ecology, 35*, 2179–90.

Rountree, R. A., & Able, K. W. (2007). Spatial and temporal habitat use patterns for salt marsh nekton: Implications for ecological functions. *Aquatic Ecology, 41*, 25–45.

Rowan, A. T. (1988). *Bank erosion, bank stability, and channel migration in salt marsh tidal channels* [Master's thesis, Rutgers University].

Shisler, J. K. (1980). Historical development of OMWM in New Jersey—equipment and techniques. *Proceedings and Papers of the Thirty-second Annual Meeting of the Utah Mosquito Abatement Association* (1979), 30–32.

Shisler, J. K, & Jobbins, D. M. (1977). Tidal variations in the movement of organic carbon in New Jersey salt marshes. *Marine Biology, 40*(2), 127–34.

Smith, J. A. M. (2013). The role of *Phragmites australis* in mediating inland salt marsh migration in a mid-Atlantic estuary. *PLoS ONE, 8*(5), e65091. doi:10.1371/journal.pone.0065091.

Smith, K. J. (1995). *Processes regulating habitat use by salt marsh nekton in a southern New Jersey estuary* [Doctoral thesis, Rutgers University].

Smith, K. J., & Able, K. W. (1994). Salt marsh tide pools as winter refuges for the mummichog, *Fundulus heteroclitus*, in New Jersey. *Estuaries, 17*(1), 226–34.

Sullivan, M. C., Wuenschel, M. J., & Able, K. W. (2009). Inter- and intra-estuary variability in ingress, condition, and settlement of the American eel *Anguilla rostrata*: implications for estimating and understanding recruitment. *Journal of Fish Biology, 74*, 1949–69.

Talbot, C. W., Able, K. W., & Shisler, J. K. (1986). Fish species composition in New Jersey salt marshes: Effects of marsh alterations for mosquito control. *Transactions of the American Fisheries Society, 115*(2), 269–78.

Tonjes, D. J. (2013). Impacts from ditching salt marshes in the mid-Atlantic and northeastern United States. *Environmental Review, 21*, 116–126.

Wimp, G. M., Murphy, S. M., Lewis, D., & Ries, L. (2011). Do edge responses cascade up or down a multi-trophic food web? *Ecology Letters, 14*, 863–70.

Wimp, G. M., Ries, L., Lewis, D., & Murphy, S. M. (2019). Habitat edge responses of generalist predators are predicted by prey and structural resources. *Ecology, 100*(6) e02662.

Wimp, G. M., Lewis, D., & Murphy, S. (2019). Impacts of nutrient subsidies on salt marsh food webs: A latitudinal survey. *Frontiers in Ecology and Evolution, 7*, 350.

Wimp, G. M., & Murphy, S. M. (2021). Habitat edges alter arthropod community composition. Landscape Ecology https://doi.org/10.1007/s10980-021-01288-6.

Chapter 8 References

Able, K. W. (1973). A new cyclopterid fish, *Liparis inquilinus*, associated with the sea scallop, *Placopecten magellanicus*, in the western North Atlantic, with notes on the *Liparis liparis coplex. Copeia, 4*, 787–94.

Able, K. W., & Musick, J. A. (1976). Life history, ecology, and behavior of *Liparis inquilinus* (Pisces Cyclopteridae) associated with the sea scallop, *Placopecten magellanicus. Fishery Bulletin, 74*(2), 409–421.

Able, K. W. (1999). Measures of juvenile fish habitat quality: Examples from a National Estuarine Research Reserve. In L. R. Benaka (Ed.), *Fish habitat: Essential fish habitat and rehabilitation* (pp. 134–47). American Fisheries Society Symposium 22.

Able, K. W., & Cowen, R. K. (2002). Seasonal distribution and abundance of lobster (*Homarus americanus*) postlarvae in the New York Bight and an adjacent estuary. *Bulletin of the New Jersey Academy of Science, 47*(2), 15–20.

Able, K. W. (2016). Natural history: An approach whole time has come, passed, and needs to be resurrected. *ICES Journal of Marine Science, 73*(9), 2150–55.

Able, K. W. (2020). *Beneath the surface: Understanding nature in the Mullica Valley Estuary*. Rutgers University Press.

Able, K. W., Smith, K. J., & Hagan, S. M. (2005). Fish composition and abundance in New Jersey marsh pools: Sampling technique effects. *Northeastern Naturalist, 12*(4), 485–502.

Able, K. W., Hales Jr., L. S., & Hagan, S. M. (2005). Movement and growth of juvenile (age 0 and 1+) tautog (*Tautoga onitis* [L.] and cunner (*Tautogolabrus adspersus* [Walbaum] in a southern New Jersey estuary. *Journal of Experimental Marine Biology and Ecology, 327*, 22–35.

Able, K. W., & Grothues, T. M. (2007). Diversity of estuarine movements of striped bass (*Morone saxatilis*): A synoptic examination of an estuarine system in southern New Jersey. *Fishery Bulletin, 105*, 426–35.

Able, K. W., Valenti, J. L., & Grothues, T. M. (2017). Fish larval supply to and within a lagoonal estuary: Multiple sources for Barnegat Bay, New Jersey. *Environmental Biology of Fishes,100*, 663–83.

Able, K. W., Grothues, T. M., & Lopez-Duarte, P. (Winter 2018–2019). Horseshoe Crabs: Ancient Migrators. *SoJourn 3*(2), 39–44.

Able, K. W., Lopez-Duarte, P., Grothues, T. M., Berry, L., Petreccs, R., Fredericks, J., Navara, C., & Hanson, A. (2019). Horshoe crabs (*Limulus polyphemus*) habitats in a small estuary and the adjacent inner continental shelf: Linkages across Mid-Atlantic Bight seascapes. *Northeastern Naturalist, 26*(1), 287–304.

Able, K. W., Grothues, T. M., Turnure, J. T., Malone, M. A., & Henkes, G. H. (2014). Dynamics of residency and egress in selected estuarine fishes: Evidence from acoustic telemetry. *Environmental Biology of Fishes, 97*(1), 91–102.

Able, K. W., Smith, J. M., & Caridad, J. F. (2015). American eel supply to an estuary and its tributaries: Spatial variation in Barnegat Bay, New Jersey. *Northeastern Naturalist, 22*(1), 53–68.

Able, K. W., Grothues, T. M., Shaw, M. J., VanMorter, S. M., Sullivan, M. C., & Ambrose, D. D. (2020). Alewife (*Alosa pseudoharengus*) spawning and nursery areas in a sentinel estuary: Spatial and temporal patterns. *Environmental Biology of Fishes, 103*(11), 1419–36.

Bertness, M. D. (1999). *The Ecology of Atlantic Shores*. Sinauer Associates Inc.

Burger, J. (1996). *A naturalist along the Jersey Shore*. Rutgers University Press.

Burger, J., & Gochfield, M. (2000). *Twenty-five nature spectacles in New Jersey*. Rutgers University Press.

Cottam, C., & Munroe, D. A. (1954). Eelgrass status and environmental relations. *Journal of Wildlife Management, 18*, 449–60.

DeRivera, C. E., Ruiz, G. M., Hines, A. H., & Jivoff, P. (2005). Biotic resistance to invasion: Native predator limits abundance and distribution of an introduced crab. *Ecology, 86*(12), 3364–76.

Easy, T. (2001). *Shore bird shooting on Sheepshead flats*. Ocean County Decoy and Gunning Show.

Epifanio, C. E., Dittel, A. I, Park, S., et al. (1998). Early life history of *Hemigrapsus sanguineus*, a non-indigenous crab in the Middle Atlantic Bight (USA). *Marine Ecology Progress Series, 170*, 231–38.

Farmer, M. A., & Triemer, R. E. (1988). A redescription of the genus Ploeotia Duj (Euglenophyceae). *Taxon 37*(2), 319–25.

Ford, S. (1947). History and present status of molluscan shellfisheries from Barnegat Bay to Delaware Bay. *NOAA Technical Report NMFS*, 127, 119–40.

Gittman, R. K., & Keller, D. A. (2013). Fiddler crabs facilitate *Spartina alterniflora* growth, mitigating periwinkle overgrazing of marsh habitat. *Ecology, 94*, 2709–18.

Gochfield, M., & Burger, J. (1997). *Butterflies of New Jersey*. Rutgers University Press.

Goodwin, J. D., Munroe, D. M., Define, Z., et al. (2019). Estimating connectivity of hard clam (*Mercenaria mercenaria*) and Eastern oyster (*Crassostrea virginica*) larvae in Barnegat Bay. *Journal of Marine Science and Engineering, 7*(167), 1–17.

Hall, J. (2017). The oyster farmers. *Edible Jersey*, 49–50.

Hall, J. (2017). How Rutgers revitalized an industry. *Edible Jersey*, 52–53.

Hall, J. (2017). Old stock, new generation. *Edible Jersey*, 45–46.

Hansens, E., & Race, S. (2008). The greenhead and you. http://www.rci.rutgers.edu/~insects/greenheads.htm.

Hoden, R. (2008). The seasonal abundance of migratory shorebirds, gulls, and geese on an intertidal mudflat in southern New Jersey: Long term observations. *Rutgers University Institute of Marine and Coastal Science Technical Report*, 97–13.

Haroski, D. M. (1998). *Fish assemblages of inlet beaches in a southern New Jersey estuary* [Masters thesis, Rutgers University].

Jacobson, N. R., Hansens, E. J., Vrijenhoek, R. C., Swofford, D. L., & Berlocher, S. H. (1981). Electrophoretic detection of a sibling species of the salt marsh greenhead, *Tabanus nigrovittatus. Annals of Entomological Society of America, 74*(6), 602–605.

Kennish, M. J. (2004). Jacques Cousteau National Estuarine Research Reserve. In M. J. Kennish (Ed.), *Estuarine research monitoring and resource protection* (pp. 59–115). CRC Press.

Kennish, M. J., Cricker, S. B, Dennison, W. C., et al. (2007). Barnegat Bay–Little Egg Harbor estuary: Case study of a highly eutrophic coastal bay system. *Ecological Applications, 17*(5), S3–S16.

Lathrop, R. G., Cole, M. B., & Showalter, R. D. (2000). Quantifying the habitat structure and spatial pattern of New Jersey (U.S.A.) salt marshes under different management regimes. *Wetlands Ecology and Management, 8*, 163–72.

Leck, C. (1984). *The status and distribution of New Jersey's birds.* Rutgers University Press.

Lees, D. C., & Zilli, A. (2021). *Moths: A complete guide to biology and behavior.* Smithsonian Books.

Mann, D. A., & Grothues, T. M. (2009). Short-term upwelling events modulate fish sound production at a mid-Atlantic Ocean observatory. *Marine Ecology Progress Series, 375*, 65–71.

McDermott, J. J. (1991). A breeding population of the western Pacific crab *Hemigrapsus sanguineaus* (Crustacea: Decapoda) established on the Atlantic Coast of North America. *Biological Bulletin, 181*, 195–98.

McKenna, K. K., DiCosmo, N., Maxwell-Doyle, M., Reilly, J., Greenfield, B., Jensen, H., & Gebert, J. (2017). *Quantification of flood event forcing and the impact of natural wetlands systems: Great Bay Boulevard, Ocean County, New Jersey*. Draft report to the Federal Highway Administration (FHWA) green infrastructure techniques for coastal highway resilience, 2016–2017 pilot program. U.S. Army Corps of Engineers Institute for Water Resources in partnership with the U.S. Army Corps of Engineers Philadelphia District, Stockton University Coastal Research Center, and Barnegat Bay Partnership.

Meixler, M. S., Kennish, M. J., & Crowley, K. F. (2017). Assessment of plant community characteristics in natural and human-altered coastal marsh ecosystems. *Estuaries and Coasts*, DOI 10.1007/s12237-017-0296-0.

Meixler, M. S., Kennish, M. J., & Sakowicz, G. P. (2018). Long-term epifaunal community assessment in estuarine waters of New Jersey. *Bulletin of New Jersey Academy Sciences, 63*(1), 9–17.

Montgomery, J. D., & Newcomb, M. R. (1975). A quantitative study of the vegetation near the Great Bay Boulevard, Tuckerton, New Jersey, vol. 3.

Ecological studies in the bays and other waterways near Little Egg Inlet and in the ocean in the vicinity of the proposed site for the Atlantic Generating Station, New Jersey. Progress report for the period January–December 1974. Ichthyological Associates, Inc.

Moul, E. T. (1985). Ecological observation of pools in the salt marshes of New Jersey. In R. A. Ragotzkie (Ed.), *ProcSalt Marsh Conference*, Marine Institute, University of Georgia Press, Athens.

Mountford, K. (1980). Occurrence and predation by *Mnemiopsis leidyi* in Barnegat Bay NJ. *Estuarine, Coastal and Shelf Science, 10*, 393–402.

Ng, C. L., Able, K. W., & Grothues, T. M. (2007). Habitat use, site fidelity, and movement of adult striped bass in a southern New Jersey estuary based on mobile acoustic telemetry. *Transactions of the American Fisheries Society, 136*, 1344–55.

Sackett, D. K., Able, K. W., & Grothues, T. M. (2007). Dynamics of summer flounder, *Paralichthys dentatus*, seasonal migration based on ultrasonic telemetry. *Estuarine, Coastal and Shelf Science, 74*, 119–30.

Schreiber, G., Lopez-Duarte, P. C., & Able, K. W. (2019). Composition, seasonality, and life history of decapod shrimps in Great Bay, New Jersey. *Northeastern Naturalist, 26*(4), 817–34.

Smith, K. J. (1995). *Processing regulating habitat use by salt marsh nekton in a southern New Jersey estuary* [Doctoral dissertation, Rutgers University].

Smith, K. J., & Able, K. W. (1994). Salt-marsh tide pools as winter refuges for the mummichog, *Fundulus heteroclitus*, in New Jersey. *Estuaries, 17*, 226–34.

Sofield, R. K., Douglas, M. E., Hansens, E. J., & Vrijenhoek, R. V. (1984). Diagnosis and detection of cryptic species: The *Tabanus nigrovittatus* complex (Diptera: Tabanidae) in coastal New Jersey. *Annals of the Entomological Society of America ,77*(5), 587–91.

Sogard, S. M. (1989). Colonization of artificial seagrass by fishes and decapod crustaceans: importance of proximity to natural eelgrass. *Journal of Experimental Marine Biology and Ecology, 133*, 15–37.

Special Report: Oysters. (2017). *Edible Jersey*.

Stone, W. (1911). *The plants of southern New Jersey*. Report of the New Jersey State Museum for 1911, 21–828.

Sullivan, M. C., Able, K. W., Hare, J. A., & Walsh, H. J. (2006). *Anguilla rostrata* glass eel ingress into two U.S. east coast estuaries; patterns, processes and implications for adult abundance. *Journal of Fish Biology, 69*, 1081–1101.

Szedlmayer, S. T., & Able, K. W. (1996). Patterns of seasonal availability and habitat use by fishes and decapod crustaceans in a southern New Jersey estuary. *Estuaries, 19*, 697–709.

Szerlag, S., & McRobert, S. P. (2006). Road occurrence and mortality of the northern diamondback terrapin. *Applied Herpetology, 3*, 27–37.

Szerlag-Egger, S., & McRobert, S. P. (2007). Northern diamondback terrapin occurrence, movement, and nesting activity along a salt marsh access road. *Chelonian Conservation and Biology, 6*(2), 295–301.

Taylor, D. L., Nichols, R., & Able, K. W. (2007). Habitat selection and quality for multiple cohorts of young of the year bluefish *Pomatomus saltatrix*: Comparisons between estuarine and ocean beaches in southern New Jersey. *Estuarine, Coastal and Shelf Science, 73*, 667–79.

Taylor, P. D., Crewe, T. L., Mackenzie, S. A., et al. (2017). The Motus Wildlife Tracking System: a collaborative research network to enhance the understanding of wildlife movement. *Avian Conservation and Ecology, 12*(1), Art 8.

Thompson, C. M., Evort, S., & Balz, J. A. (2022). Comparing oyster population structure and disease prevalence between a disease-resistant and transplanted wild population in Barnegat Bay, NJ (U.S.A.). *Restoration Ecology, 30*(3), e13628.

Toth, J. L., Hohn, A. A., Able, K. W., & Gorgone, A. M. (2011). Defining bottlenose dolphin (*Tursiops truncatus*) population units based on environmental, physical, and behavioral characteristics. *Marine Mammal Science, 27*(1), 94–110.

Toth, J. L., Hohn, A. A., Able, K. W., & Gorgone, A. M. (2011). Patterns of seasonal occurrence, distribution and site fidelity of coastal bottlenose dolphins (*Tursiops truncatus*) in southern New Jersey, U.S.A. *Marine Mammal Science, 27*(1), 94–110.

Toth, J., Evert, S., Zimmermann, E., Sullivan, M., Dotts, L., Able, K. W., Hagan, R., & Slocum, C. (2018). Annual residency patterns and diet of *Phoca vitulina concolor* (West Atlantic Harbor Seal) in a southern New Jersey estuary. *Northeastern Naturalist, 25*(4), 611–26.

Triemer, R. E. (1986). Light and electron microscopic description of a colorless euglenoid, *Serpenomonas costata* n.g., n.sp. *Journal of Protozoology, 33*, 412–15.

Turnure, J. T., Able, K. W., & Grothues, T. M. (2015). Seasonal residence of adult weakfish (*Cynoscion regalis*) in a small temperate estuary based on acoustic telemetry: A local perspective of a coast wide phenomenon. *Environmental Biology of Fishes, 98*, 1207–21.

Turnure, J. T., Able, K. W., & Grothues, T. M. (2015). Patterns of intra-estuarine movement in adult weakfish *(Cynoscion regalis*): evidence of site-affinity at seasonal and diel scales. *Fishery Bulletin, 113*, 167–79.

Viscido, S. V., Stearns, D. E., & Able, K. W. (1997). Seasonal and spatial patterns of an epibenthic decapod crustacean assemblage in north-west Atlantic continental shelf waters. *Estuarine, Coastal and Shelf Science, 45*, 377–92.

Waddell, E. C., Piniak, W. E. D., Reinsel, K. A., & Welch, J. M. (2021). The role of adult fiddler crab environmental acoustic cues and chemical cues in stimulating molting of field-caught megalopae. *Journal of Experimental Marine Biology and Ecology, 537*, 151516.

Wang, J. Q., Zhang, X. D., Jiang, L. F., et al. (2010). Bioturbation of burrowing crabs promotes sediment turnover and carbon and nitrogen movements in an estuarine salt marsh. *Ecosystems, 13*, 586–99.

Wilson, K. A., Able, K. W., & Heck, K. L. (1990). Habitat use by juvenile blue crabs: a comparison among habitats in southern New Jersey. *Bulletin of Marine Science, 46*, 105–114.

Wilson, K. A., Able, K. W., & Heck Jr., K. L. (1990). Predation rates on juvenile blue crabs in estuarine nursery habitats: Evidence for the importance of macroalgae (*Ulva lactuca*). *Marine Ecology Progress Series, 58*, 243–51.

Windham, L., & Lathrop, R. (1999). Effects of *Phragmites australis* (common reed) invasion on above-ground biomass and soil properties in brackish tidal marsh of the Mullica River, New Jersey. *Estuaries, 22*, 927–35.

Woodland, R. J., Rowe, C. L., & Henry, P. F. P. (2017). Changes in habitat availability for multiple life stages of diamondback terrapins (*Malaclemys terrapin*) in Chesapeake Bay in response to sea level rise. *Estuaries and Coasts, 4*(1), 78–80.

Wuenschel, M. J., & Able, K. W. (2008). Swimming ability of eels (*Anguilla rostrata, Conger oceanicus*) at estuarine ingress: Contrasting patterns of cross-shelf transport. *Marine Biology, 154*, 775–86.

Young, T. P. (2000). Restoration ecology and conservation biology. *Biological Conservation Journal, 92*, 73–83.

Chapter 9 References

Able, K. W. (2015). *Station 119: From lifesaving to marine research*. Down the Shore Publishing.

Able, K. W. (2016). Natural history: An approach whose time has come, passed, and needs to be resurrected. *ICES Journal of Marine Science*, doi:10.1093/icesjms/fsw049.

Able, K. W. (2020). Diamondback terrapins and the lost years. In *Beneath the surface: Understanding nature in the Mullica Valley Estuary* (pp. 222–23). Rutgers University Press.

Avissar, N. G. (2006). Changes in population structure of diamondback terrapins (*Malaclemys terrapin terrapin*) in a previously surveyed creek in southern New Jersey. *Chelonian Conservation and Biology, 5*(1), 154–59.

Brennessel, B. (2006). *Diamonds in the marsh: A natural history of the Diamondback Terrapin*. University Press of New England.

Burger, J. (1982). Diamondback terrapins: Between laying and hatching. *New Jersey Outdoors, 9*(4), 2–3, 30–31.

Burger, J. (1996). *A naturalist along the shore*. Rutgers University Press.

Forman, T. T. T, Sperling, D., Bissonette, J. A., et al. (2003). *Road ecology: Science and solutions*. Island Press.

Hansley, T. C., Kimbro, D. L., & Hughes, A. R. (2017). Stress and subsidy effects of seagrass wrack duration, frequency, and magnitude on salt marsh community structure. *Ecology, 98*(7), 1884–95.

Hartman, J., Caswell, H., & Valiela, I. (1983). Effects of wrack accumulation on salt marsh vegetation. *Oceanologica Acta Proceedings*, 17th European Marine Biology Symposium, Brest, France, 27 September–1 October 1982, 99–102.

Hoden, R., & Able, K. W. (2003). Habitats use and road kill mortality of diamondback terrapin (*Malaclemys terrapin*) in the Jacques Cousteau National Estuarine Research Reserve at Mullica River–Great Bay in southern New Jersey. Jacques Cousteau NERR Technical Report, 100–24.

Michinton, T. E. (2002). Disturbance by wrack facilitates spread of Phragmites australis in a coastal marsh. *Journal of Experimental Marine Biology and Ecology, 281*, 89–107.

Montgomery, J. D., & Newcomb, M. R. (1975). A quantitative study of the vegetation near the Great Bay Boulevard, Tuckerton, New Jersey, vol. 3. *Ecological studies in the bays and other waterways near Little Egg Inlet and in the ocean in the vicinity of the proposed site for the Atlantic Generating Station, New Jersey. Progress report for the period January–December 1974.* Ichthyological Associates, Inc.

Rippel, T. M., Tomasula, J., Murphy, S. M., & Wimp, G. W. (2021). Global change in marine coastal habitats impacts insect populations and communities. *Current Opinions in Insect Science, 47*, 1–6.

Sheridan, C. M., Spotila, J. R., Bien, W. F., & Avery, H. W. (2010). Sex-biased dispersal and natal philopatry in the diamondback terrapin, *Malaclemys terrapin. Molecular Ecology, 19*, 5497–5510.

Szerlag, S., & McRobert, S. P. (2006). Road occurrence and mortality of the northern diamondback terrapin. *Applied Herpetology, 3*(1), 27–37.

Szerlag-Egger, S., & McRobert, S. P. (2007). Northern diamondback terrapin occurrence, movement, and nesting activity along a salt marsh access road. *Chelonian Conservation and Biology*, 295–301.

Woodland, R. J., Rowe, C. L., & Henry, P. F. P. (2017). Changes in habitat availability for multiple life stages of Diamondback Terrapins (*Malaclemys terrapin*) in Chesapeake Bay in response to Sea Level Rise. *Estuaries and Coasts, 40*, 1502–1515.

Chapter 10 References

Able, K. W., & Rosen, M. (2020). Impacts of thin layer deposition on salt marsh vegetation and fauna. Technical Report, Rutgers University Marine Field Station.

Able, K. W., Grothues, T. M., Morson, J. M., & Coleman, K. E. (2014). Temporal variation in winter flounder recruitment at the southern margin of their range: Is the decline due to increasing temperatures? *ICES Journal of Marine Science, 71*(8), 2186–91.

Able, K. W. (2020). *Beneath the surface: Understanding nature in the Mullica Valley Estuary*. Rutgers University Press.

Able, K. W. (2022). From cedar cemeteries to marsh lakes: A case study of sea-level rise and habitat change in a northeastern US salt marsh. *Estuaries and Coasts, 44*, 1649–57.

Adamowicz, S. C., Wilson, G., Burdick, D. M., Ferguson, W., & Hopping, R. (2020). Farmers in the marsh: Lessons from history and case studies for the future. *Wetlands Science and Practice, 20*, 183–94.

Baker, R., Taylor, M. D., Able, K. W., et al. (2020). Fisheries rely on threatened salt marshes. *Science, 370*, 670–71.

Baum, J. K., & Myers, R. A. (2004). Shifting baselines and the decline of pelagic sharks in the Gulf of Mexico. *Ecology Letters, 7*, 135–45.

Bell, R. J., Richardson, D. E., Hare J. A., Lynch, P. D., & Fratantoni, P. S. (2015). Disentangling the effects of climate, abundance, and size on the distribution of marine fish: An example based on four stocks from the Northeast US shelf. *ICES Journal of Marine Science, 72*(5), 1311–22.

Bonebrake, T. C., Christensen, J., Boggs, C. L., & Ehrlich, P. R. (2010). Population decline assessment, historical baselines, and conservation. *Conservation Letters, 3*, 371–78.

Borchert, S. M., Osland, M. J., Enwright, N. M., & Griffith, K. T. (2018). Coastal wetland adaptation to sea level rise: Quantifying potential for landward migration and coastal squeeze. *Journal of Applied Ecology, 55*, 2876–87.

Burns, C. J., Alber, M., & Alexander, C. R. (2021). Historical changes in the vegetated area of salt marshes. *Estuaries and Coasts, 44*, 162–77.

Cooke, S. J., Auld, H. L., Birnie-Gauvin, K., et al. (2022). On the relevance of animal behavior to the management and conservation of fishes and fisheries. *Environmental Biology of Fishes, 106*, 785–810.

Donnelly, J. P., Butler, J., Roll, S., et al. (2004). A backbarrier overwash record of intense storms from Brigantine, New Jersey. *Marine Geology, 210*, 107–21.

Ganju, N. K. (2019). Marshes are the new beaches: Integrating sediment transport into restoration planning. *Estuaries and Coasts, 42*, 917–26.

Gaul, G. M. (2019). *The geography of risk: Epic storms, rising seas, and the cost of America's coasts*. Sarah Crichton Books.

Grothues, T. M., Morson, J. M., Hagan, R., & Able, K. W. (In prep). Sensitivity of estuarine larval and juvenile fish and crab phenology to multiple benchmarks in a twenty-five year time series.

Haaf, L., Watson, E. B., Elsey-Quirk, T., et al. (2022). Sediment accumulation, elevation change, and the vulnerability of tidal marshes in the Delaware Estuary and Barnegat Bay to accelerated sea level rise. *Estuaries and Coasts*, doi: 10.1007/s12237-021-00972-9.

Harper, R. M. (1912). The diverse habitats of Eastern red cedar and their interpretation. *Torreya, 12*(7), 145–54.

Jackson, J. B. C., Alexander, K. E., & Sala, E. (2011). *Shifting baselines: The past and future of ocean fisheries*. Island Press.

Jackson, J. B. C., Kirby, M. X., Berger, W. H., Bjorndal, K. A., Botsford, L. W., et al. (2001). Historical overfishing and the recent collapse of coastal ecosystems. *Science, 293*, 629–38.

Kennish, M. J., Spahn, A., & Sakawicz, G. P. (2014a). Sentinel site development of a major salt marsh system in the Mid-Atlantic region (USA). *Open Journal of Ecology, 4*, 77–86.

Kennish, M. J., Meixler, M. S., Petruzelli, G., & Fertig, B. (2014b). Tuckerton Peninsula salt marsh system: A sentinel site for assessing climate change effects. Bull. *New Jersey Academy Science, 59*(2), 1–5.

Kennish, M. J., Lathrop Jr., R. G., Spahn, A., Sakowicz, G. P., & Scatelli, R. (2016). The JCNERR Sentinel Site: Research and monitoring applications. *Bulletin New Jersey Academy of Science, 61*(1), 1–8.

Kennish, M. J., Paerl, N. W., & Crosswell, J. R. (2023). *Climate change and estuaries*. CRC Press.

Klingbeil, B. T., Cohen, J. B., Correll, M. D., et al. (2021). High uncertainty over the future of tidal marsh birds under current sea-level rise projections. *Biodiversity and Conservation, 30*, 431–43.

Knowlton, N., & Jackson, J. B. C. (2008). Shifting baselines, local impacts, and global change on coral reefs. *PLos Biology, 6*(2), 0215–0220.

Lauchalan, S. S., & Nagelkerken, I. (2020). Species range shifts along multistressor mosaics in estuarine environments under future climate. *Fish*

and Fisheries, 21, 32–46.
Leo, K. L., Gilles, C. L., Fitzsimons, J. A., Hale, L. Z., & Beck, M. W. (2019). Coastal habitat squeeze: A review of adaptation solutions for saltmarsh, mangrove and beach habitats. *Ocean and Coastal Management, 175*, 180–90.
Linkhurst, R. A., & Seneca, E. D. (1980). The effects of standing water and drainage potential on the *Spartina alterniflora*-substrate complex in a North Carolina salt marsh. *Estuarine and Coastal Marine Science, 11*, 41–52.
Lotze, H. K., & Worm, B. (2008). Historical baselines for large marine animals. *Trends in Ecology and Evolution, 24*(5), 254–62.
Mariotti, G. (2016). Revisiting salt marsh resilience to sea level rise: Are ponds responsible for permanent land loss? *Journal of Geophysical Research: Earth Surface, 12*, 1391–1407.
Marriotti, G. (2020). Beyond marsh drowning: The many faces of marsh loss (and gain). *Advances in Water Resources, 144*, 103710.
Mazzagetti, D. (2018). *The Jersey Shore: The past, present and future of a National Treasure*. Rutgers University Press.
Mazzella, S. (2013). *Surviving Sandy: Long Beach Island and the greatest storm of the Jersey Shore*. Down the Shore Publishing.
McBride, R. S., & Able, K. W. (1998). Ecology and fate of butterflyfishes, *Chaetodon spp*., in the temperate, western North Atlantic. *Bulletin of Marine Science, 63*(2), 401–16.
McClenachan, L., Ferretti, F., & Baum, J. K. (2012). From archives to conservation: Why historical data are needed to set baselines for marine animals and ecosystems. *Conservation Letters, 5*, 349–59.
Miller, S. H., Breitburg, D. L., Burrell, R. B., & Keppel, A. G. (2016). Acidification increases sensitivity to hypoxia in important forage fishes. *Marine Ecology Progress Series, 549*, 1–8.
Morley, J. W., Selden, R. L., Latour, R. J., Frolicher, T. L., Seagraves, R. J., & Pinsky, M. L. (2018). Projecting shifts in thermal habitat for 686 species on the North American continental shelf. *PLOS ONE, 13*(5), e0196127.
Morson, J. M., Grothues, T., & Able, K. W. (2019). Change in larval fish assemblage in a USA east coast estuary estimated from twenty-six years of fixed weekly sampling. *PLOS ONE, 14*(10), e0224157.
Mounier, R. A. (1999). A brief history of investigation of the Tuckerton Shell Mound. *Bulletin of the Archaeological Society of New Jersey, 54*, 43–44.
Mounier, R. A. (2004). The Tuckerton Shell Mound: A lasting fertilizer. Paper presented at *From the Arctic to Avalon: Transforming the History of Northeastern North America*, St. John's Newfoundland.
Nickerson, K. J. (2017). *Changing fish assemblages in a Mid-Atlantic estuary:*

Sensitivity analysis of a long term time series to climate change and guild construction [Masters thesis, Rutgers University].

Nickerson, K. J., Grothues, T. M., & Able, K. W. (2018). Sensitivity of a fish time-series analysis to guild construction: A case study of the Mullica River–Great Bay ecosystem. *Marine Ecology Progress Series, 598*, 113–29.

NOAA. (2022). National Centers for Environmental Information, State climate summaries–New Jersey.

Nordstrom, K. F., & Roman, C. T. (1996). *Estuarine shores: Evolution, environments and human alterations.* John Wiley and Sons.

Nordstrom, K. F., Jackson, N. L., & Roman, C. T. (2016). Facilitating landform migration by removing shore protection structures: opportunities and constraints. *Environmental Science and Policy, 66*, 217–26.

Nye, J. A., Link, J. S., Hare, J. A., & Ovreholtz, W. J. (2009). Changing spatial distribution of fish stocks in relation to climate and population size on the Northeast United States continental shelf. *Marine Ecology Progress Series, 393*, 111–29.

Ohenhen, L. O., Shirzaei, M., & Barnard, P. L. (2024). Slowly but surely: Exposure of communities and infrastructure to subsidence on the US east coast. *PNAS Nexus, 3*, 1–14.

Olson, E., & Vasslides, J. M. (2022). Multi-decadal declines and species assemblage shifts in the fish community of a northeast US temperate estuary. *Estuaries and Coasts, 45*, 22–40.

Papworth, S. K., Rist, J., Coad, L., Milner-Gulland, E. J. (2009). Evidence for shifting baseline syndrome in conservation. *Conservation Letters*, 2, 93–100.

Pauly, D. (1995). Anecdotes and the shifting baseline syndrome of fisheries. *Trends Ecology and Evolution, 10*, 430.

Peterson, C. H., Able, K. W., DeJong, C. F., et al. (2008). Practical proxies for tidal marsh ecosystem services: Application to injury and restoration. In D. W. Sims (ed.), *Advances in Marine Biology*, 54, Academic Press, Elsevier Inc., 221–66.

Pinsky, M. L., Worm, B., Fogarty, M. J., et al. (2013). Marine taxa track local climate velocities. *Science, 341*, 1239–42.

Pinnegar, J. K., & Engelhard, G. H. (2007). The "shifting baseline" phenomenon: A global perspective. *Reviews in Fish Biology and Fisheries, 18*, 1–16.

Raposa, K. B., Woolfolk, A., Endris, C. A., et al. (2023). Evaluating thin-layer sediment placement as a tool for enhancing tidal marsh resilience: A coordinated experiment across eight U.S. National Estuarine Research Reserves. *Estuaries and Coasts, 46*, 595–615.

Rippel, T. M., Tomasula, J., Murphy, S. M., & Wimp, G. W. (2021). Global change in marine coastal habitats impacts insect populations and communities. *Current Opinions in Insect Science, 47*, 1–6.

Rippel, T. M., Minsavage-Davis, C. D., Shirey, V., & Wimp, G. M. (2023). Simple machine learning with aerial imagery reveals severe loss of a salt marsh foundation species. *Estuaries and Coasts, 46*, 1110–22.

Sacatelli, R., Kaplan, M., Carleton, G., & Lathrop, R. G. (2023). Coastal forest dieback in the Northeast USA: Potential mechanisms and management responses. *Sustainability, 15*(8), 6346.

Saenz-Arroyo, A., Roberts, C. M., Torre, J., Carino-Olvera, M., & Enriquez-Andrade, R. R. (2005). Rapidly shifting environmental baselines among fishers of the Gulf of California. *Proceedings of the Royal Society B, 272*, 1957–62.

Saintilan, N., Kovalenko, K. E., Guntenspergen, G., et al. (2022). Constraints on the adjustment of tidal marshes to accelerating sea level rise. *Science, 377*, 523–27.

Schepers, L. P., Kirwan, M., Guntenspergen, G., & Temmerman, S. (2016). Spatio-temporal development of vegetation die-off in a submerging coastal marsh. *Limnology and Oceanography, 62*, 137–50.

Schepers, L. P., Brennand, P., Kirwin, M. L., et al. (2020). Coastal marsh degradation into ponds induces irreversible elevation loss relative to sea level in a microtidal system. *Geophysical Research Letters, 47*, e2020GL089121.

Schile, L. M., Callaway, J. C., Morris, J. T., et al. (2014). Modeling tidal marsh distribution with sea-level rise: Evaluating the role of vegetation, sediment, and upland habitat in marsh resiliency. *PLoS One, 9*(2), e88760. doi: 10.1371/journal.pone.0088760.

Schuerch, M. T., Spencer, T., Temmerman, S., et al. (2018). Future response of global coastal wetlands to sea-level rise. *Nature Research Letter, 561*, 231–46.

Shackel., N. L., Ricard, D., & Stortini, C. (2014). Thermal habitat index of many northwest Atlantic temperate species stays neutral under warming projected for 2030 but changes radically by 2060. *PLoS One, 9*(3), e90662. doi: 10.1371/journal.pone.0090662.

Sheridan, C. M., Spotila, J. R., Bien, W. F., and Avery, H. W. (2010). Sex-biased dispersal and natal philopatry in the diamondback terrapin, *Malaclemys terrapin. Molecular Ecology, 19*, 5497–5510.

Smith, J. A. M. (2013). The role of *Phragmites australis* in mediating inland salt marsh migration in a Mid-Atlantic estuary. *PLOS One, 8*(5), P65091.

Smith, J. A. M., & Pellew, M. (2021). Pond dynamics yield minimal net loss of vegetation cover across an unditched salt marsh landscape. *Estuaries and Coasts, 44*, 1534–46.

Smith, S. M. (2009). Multi-decadal changes in salt marshes of Cape Cod, MA: Photographic analyses of vegetation loss, species shifts, and geomorphic change. *Northeastern Naturalist, 16*(2), 183–208.

Stanzeski, A. (2001). The Tuckerton Shell Mound. *Bulletin of the Archaeological Society of New Jersey, 56*, 47–50.

Suk, N. S., Guo, Q., & Psuty, N. P. (1999). Suspended solids flux between salt marsh and adjacent bay: A long-term continuous measurement. *Estuaries, Coasts, and Shelf Science, 49*, 61–81.

Sweet, W., Park, J., Marra, J., Zervas, C., & Gill, S. (2014). Sea level rise and nuisance flood frequency changes around the United States. *NOAA Technical Report* NOS CO-OPS 073.

Todd, R. E., & Ren, A. S. (2023). Warming and lateral shift of the Gulf Stream from in situ observations since 2001. *Nature Climate Change*. DOI: 10.1038/s41558-023-01835-w.

Tolliver, K. W., Martin, D. W., & Young, D. R. (1997). Freshwater and saltwater flooding response for woody species common to barrier island swales. *Wetlands, 17*(1), 10–18.

U.S. Army Corps of Engineers (Philadelphia District). (2024). New Jersey back bays coastal storm risk management: Supplemental draft integrated feasibility report and environmental impact statement. E15X-202-00-EFP-1734174725.

Valenti, J. L., Grothues, T. M., & Able, K. W. (2020). Juvenile fish assemblage recruitment dynamics in a mid-Atlantic estuary: Before and after Hurricane Sandy. *Marine Ecology Progress Series, 641*, 177–93.

Vinent, O. D., Herbert, E. R., Coleman, D. J., Himmelstein, J. D., & Kirwan, M. L. (2021). Onset of runaway fragmentation of salt marshes. *One Earth, 4*, 506–516.

Walker, J. S., Kopp, R. E, Shaw, T. A., et al. (2021). Common Era sea-level budgets along the U.S. Atlantic Coast. *Nature Communications, 12*, 1841. DOI: doi.org/10.1038/s41467-021-22079-2.

Weis, J. S., Watson, E. B., Ravit, B., Harman, C., & Yepsen, M. (2021). The status and future of tidal marshes in New Jersey faced with sea level rise. *Anthropocene Coasts, 4*, 168–92.

Whitfield, A. K., Gillanders, B. M., & Able, K. W. (2023). Climate change effects on fish populations. In M. J. Kennish, H. W. Paerl, and J. R. Crosswell (Eds.), *Climate Change and Estuaries*. CRC Press.

Whitfield, A. K., Able, K. W., Barletta, M., et al. (2023). Life-history guilds of fishes associated with estuaries: Opportunism versus dependency. *Estuarine, Coastal and Shelf Science, 292*, 108456.

Wimp, G. M., Murphy, S. M., Lewis, D., & Ries, L. (2011). Do edge responses cascade up or down a multi-trophic food web? *Ecology Letters, 14*, 863–70.

Wimp, G. M., Murphy, S. M. (2021). Habitat edges alter arthropod community composition. *Landscape Ecology, 36*, 2849–2861.

Wimp, G. M., Ries, L., Lewis, D., & Murphy, S. M. (2019). Habitat edge responses of generalist predators are predicted by prey and structural resources. *Ecology, 100*, e02662. https://doi.org/10.1002/eey.2662.

Wimp, G. M., Lewis, D., Murphy, S. (2019). Impacts of nutrient subsidies on salt marsh food webs: A latitudinal survey. *Frontiers in Ecology and Evolution, 7*, 350.

Woodland, R. J., Rowe, C. L., & Henry, P. F. P. (2017). Changes in habitat availability for multiple life stages of diamondback terrapins (*Malaclemys terrapin*) in Chesapeake Bay in response to sea level rise. *Estuaries and Coast, 40*, 1502–15.

Yates, J. (2024). Hatfield's and Willet's stores/hotels on Sheepshead Meadows: Dancing with Leah Blackman's sister-in-law. *SoJourn 8*, 63–87.

Index

About the Author

Ken Able is a former Rutgers University Distinguished Professor in the Department of Marine and Coastal Sciences. As director of the Marine Field Station, he has spent over thirty years trying to understand how animals use the estuarine waters of the Mullica Valley as well as other estuaries along the east coast of the United States and the Gulf of Mexico. His diverse interests as a distinguished professor emeritus include estuarine ecosystems with a focus on life history and the ecology of fishes with an emphasis on habitat quality.

Colophon

This book has been designed and set in type by the South Jersey Culture & History Center and printed by Lightning Source, a subsidiary of Ingram Content Group. The text is composed in 12 point Minion Pro, body text, and 18 point Adelle Sans, chapter headings. Ken Able has been very patient as our staff worked to thoughtfully edit, design, and compose this book. It is a pleasure to work with you, Ken.

The mission of the South Jersey Culture & History Center is to deepen public awareness and understanding of southern New Jersey's rich cultural and historical heritage, to encourage its study, particularly among students, and to produce publications that offer lasting insight into the region.

www.ingramcontent.com/pod-product-compliance
Lightning Source LLC
Chambersburg PA
CBHW041258210926
51847CB00009B/63
9781947889347